Advances in 21st Century Human Settlements

Indexed by SCOPUS

This Series focuses on the entire spectrum of human settlements—from rural to urban, in different regions of the world, with questions such as: What factors cause and guide the process of change in human settlements from rural to urban in character, from hamlets and villages to towns, cities and megacities? Is this process different across time and space, how and why? Is there a future for rural life? Is it possible or not to have industrial development in rural settlements, and how? Why does 'urban shrinkage' occur? Are the rural areas urbanizing or is that urban areas are undergoing 'ruralisation' (in form of underserviced slums)? What are the challenges faced by 'mega urban regions', and how they can be/are being addressed? What drives economic dynamism in human settlements? Is the urban-based economic growth paradigm the only answer to the quest for sustainable development, or is there an urgent need to balance between economic growth on one hand and ecosystem restoration and conservation on the other—for the future sustainability of human habitats? How and what new technology is helping to achieve sustainable development in human settlements? What sort of changes in the current planning, management and governance of human settlements are needed to face the changing environment including the climate and increasing disaster risks? What is the uniqueness of the new 'socio-cultural spaces' that emerge in human settlements, and how they change over time? As rural settlements become urban, are the new 'urban spaces' resulting in the loss of rural life and 'socio-cultural spaces'? What is leading the preservation of rural 'socio-cultural spaces' within the urbanizing world, and how? What is the emerging nature of the rural-urban interface, and what factors influence it? What are the emerging perspectives that help understand the human-environment-culture complex through the study of human settlements and the related ecosystems, and how do they transform our understanding of cultural landscapes and 'waterscapes' in the 21st Century? What else is and/or likely to be new vis-à-vis human settlements—now and in the future? The Series, therefore, welcomes contributions with fresh cognitive perspectives to understand the new and emerging realities of the 21st Century human settlements. Such perspectives will include a multidisciplinary analysis, constituting of the demographic, spatio-economic, environmental, technological, and planning, management and governance lenses.

If you are interested in submitting a proposal for this series, please contact the Series Editor, or the Publishing Editor:

Bharat Dahiya (bharatdahiya@gmail.com) or
Loyola DSilva (loyola.dsilva@springer.com)

More information about this series at http://www.springer.com/series/13196

Chan-Hoong Leong · Lai-Choo Malone-Lee
Editors

Building Resilient Neighbourhoods in Singapore

The Convergence of Policies, Research and Practice

Springer

Editors
Chan-Hoong Leong
Institute of Policy Studies
National University of Singapore
Singapore, Singapore

Lai-Choo Malone-Lee
Centre for Sustainable Asian Cities
National University of Singapore
Singapore, Singapore

ISSN 2198-2546 ISSN 2198-2554 (electronic)
Advances in 21st Century Human Settlements
ISBN 978-981-13-7047-2 ISBN 978-981-13-7048-9 (eBook)
https://doi.org/10.1007/978-981-13-7048-9

Library of Congress Control Number: 2019933383

This Springer imprint is published by the registered company Springer Nature Singapore Pte Ltd.
The registered company address is: 152 Beach Road, #21-01/04 Gateway East, Singapore 189721, Singapore

Foreword

Singapore has always taken a long-term view towards urban planning. Through thoughtful planning of our infrastructure and our housing estates and proactive greening of our surroundings, we have a city that is vibrant and robust. We must, however, always look forward and be prepared for the challenges ahead, and build greater resilience in our social and urban fabric.

Around the world, we have seen how countries, cities and communities bounce back from natural and man-made disasters. For example, Japan, which is prone to natural disasters, continually takes steps to strengthen its infrastructure to cope with severe events. But building resilience extends beyond ensuring that our physical infrastructure and emergency plans are robust. It is also about strengthening ties among our people and developing trust in our communities and neighbourhoods, to ensure that we collectively pull together in times of crisis.

As highlighted in some of the essays in this book, building cohesive communities has always played an important role in Singapore's urban planning. For example, the People's Association was set up in 1960 to encourage the formation of community networks through its social programmes. To promote racial integration in our neighbourhoods and prevent the formation of racial enclaves, the Ethnic Integration Policy has been in place in our public housing estates since 1989. More recently, we have provided more open green spaces and opened up neighbourhood venues to hold arts and cultural programmes to further encourage community bonding.

Despite a strong foundation, we continue to learn, mature and grow, to build a more resilient Singapore. This book suggests how more resilient neighbourhoods can be built in Singapore, in the context of our dense housing estates. The essays look at resilience from different perspectives—from design and planning, policy design and implementation, to organisations and residents.

The trust, networks and support systems built over the years will be critical for Singapore in the face of such challenges ahead. We must strive to bring about greater social cohesiveness and inclusiveness, and make Singapore a more resilient, sustainable and liveable home for all Singaporeans.

Singapore

Desmond Lee
Minister for Social and Family Development

Second Minister for National Development

Contents

The Ecology of Neighbourhood Resilience: A Multi-disciplinary Perspective

Lai-Choo Malone-Lee

Abstract Policy makers and planners increasingly recognise the importance of social resilience in driving collective response and action, and build community capacity to cope with and overcome adversity when they arise. It is also recognised that social resilience involve consistent efforts to build stronger and more cohesive communities in themselves, not only to respond to adversities and warding off future threats, but also as a means to strengthening social ties and addressing inequities that may exist for vulnerable or marginalized groups. This is in line with the aspirations of the New Urban Agenda which was adopted as a shared vision by the global community for a better and more sustainable future—one in which the tenets of inclusiveness, equality, participatory involvement and strong governance are vigorously promoted. The essays in this book contribute to the on-going discourse on these thematic areas and, using Singapore as the contextual backdrop, highlight how cities can embrace this new urban paradigm in the way they are planned, designed, developed, governed and managed.

Keywords Social resilience · Neighbourhood · Planning and design · Community · Participation · Inclusiveness

1 Introduction

The emergence of resilience planning as a distinct disciplinary field is recent, and has opened up many opportunities for rich multi-disciplinary research to support policy and practice. Resiliency, with reference to the ability to bounce back, maintain essential functions, and thrive despite severe stresses or shocks (IPCC 2012; Timmerman 1981; UNISDR 2009) has been studied from various perspectives, straddling the psychological, social, ecological, engineering, infrastructural as well as the economic domains (Chelleri and Olazabal 2012). However, in the context of complex urban

L.-C. Malone-Lee (✉)
National University of Singapore, Singapore, Singapore
e-mail: sdemalon@nus.edu.sg

C.-H. Leong and L.-C. Malone-Lee (eds.), *Building Resilient Neighbourhoods in Singapore*, Advances in 21st Century Human Settlements,
https://doi.org/10.1007/978-981-13-7048-9_1

environments, the dominant threads of social discourse seem to have converged on the environmental and social-psychological realms, which are clearly interlinked (Folke et al. 2010; Resilience Alliance 2009; Walker et al. 2009). In cities, as a response to the numerous episodic occurrences of extreme climate events, infectious disease pandemics, natural or man-made disasters, economic shocks and political upheavals, resilience planning has become a key constituent of social and urban policies as well as development programs and urban plans.

Policy makers and planners increasingly recognise the importance of collective response and action, and the necessity to build community capacity to cope with and to overcome adversity when they arise. Specifically, social or community resilience is understood to encompass the triad qualities of resistance, recovery and creativity. *Resistance* relates to a community's efforts to withstand a disturbance and its consequences, and *recovery* relates to the community's ability to pull through the disturbance. The third quality, *creativity*, augments the entire experiential process, and is represented by a gain in resilience achieved as part of recovery process, which can be attained by learning through the disturbance experience (Kimhi and Shamai 2004). Building social or community resilience would primarily involve developing high levels of social capital, that is, mutual trust, defined social norms, participation and social networks (Australian Social Inclusion Board 2009) that will bring forth the ability, strength and resolve to attain these three qualities.

Social resilience, however, does not relate to crisis planning alone. It is increasingly recognised to involve consistent efforts to build stronger and more cohesive communities in themselves, not only to respond to adversities and warding off future threats. This means a focus on strengthening social ties, addressing inequities that exist for vulnerable or marginalized groups (The Community Social Planning Council of Greater Victoria 2013) developing common values and promoting inclusiveness. It is an on-going process of learning and adaptation, such that, with consistent efforts over time, through strong social policies, supported by thoughtful urban planning and environmental design, the social-ecological system gradually becomes capable of self-organizing (Carpenter et al. 2001; Folke et al. 2002; Resilience Alliance 2009).

Thus, while the primary purpose in resilience planning is to prepare and enhance a community's ability to respond pro-actively and adapt positively under crisis, the concurrent aspiration is to reap tangible outcomes in terms of enhanced social wellbeing and strengthened social ties. For the community, it means pursuing the complementary purpose of wellbeing promotion in the context of stressful urban life, while for the individual, it embodies a hope that the sense of belonging to a community would pave the way for personal development and self-actualisation.

At the international level, the New Urban Agenda (NUA) which was adopted at the United Nations Conference on Housing and Sustainable Urban Development (Habitat III) in Quito, Ecuador, in October 2016, presented a shared vision for a better and more sustainable future—one in which the tenets of inclusiveness, equality, participatory involvement and strong governance (amongst others) are vigorously promoted (United Nations 2017). These are also essential underpinnings of social resiliency. In embracing this new paradigm, cities could work toward reducing inequalities, promoting sustained, inclusive and sustainable economic growth, improving human health and wellbeing, while also fostering social resilience.

This chapter presents an introduction to the essays as set forth in this book, and highlights the way they could contribute to the on-going discourse on how cities can embrace this new urban paradigm in the way they are planned, designed, developed, governed and managed. The essays will include discussions within the broad domain of urban and territorial planning, covering the thematic subjects of ageing, the arts, green spaces and mobility, and participatory involvement in planning and design. Although much of the discussions are presented with Singapore as the contextual reference point, the enactments serve to illustrate succinctly what can be done to enhance both content and process, starting from the neighbourhood level and escalating to the national level. More importantly, while specific case-based studies are referred to, the desired qualities of belongingness, safety, accessibility, social and intergenerational interaction, cultural engagement and political participation, are universal, as also promulgated by the NUA, and they should find application in all peace-seeking and pluralistic societies.

2 Neighbourhood as the Spatial Frame for Resilience Building

There is an essential spatial-temporal aspect to resilience building. Within the bounds of time and space, resiliency in the broadest sense, based on adaptation and transformation of urban systems over a period of time, would require responses at the macro-level such as the city or even the country. On the other hand, the qualities that are needed for gradual and purposeful development of social relationships, would entail sustained local level commitment. Here, a more appropriate level of engagement would be the meso-level, within the context of a locale-based unit, as in a neighbourhood.

Planners have traditionally viewed the neighbourhood as a physical planning construct based on the twin concepts of convenience and accessibility. Planning pioneer, Clarence Perry, conceptualized the ideal neighbourhood as one centred on a primary school, with residents located within a walkable ¼ mile radius, and shops, parks and other amenities around it (Perry 1929). This conceptualization of a neighbourhood unit has, for a long time, been the foundational element for worthy town plans, and has been applied quite extensively in the planning of new towns in Singapore. While planners have harnessed this spatial concept as an organising principle in the planning and provision of local amenities such as schools, shops, markets, and the like, they have always hoped that the neighbourhood of convenience would eventually evolve into a neighbourhood for community life, one built upon mutual trust, inter-dependency and participatory engagement of involved citizens. However, the reality is that while urban planners and designers have hoped that by creating places that enhance the opportunity for encounters and meaningful interactions, such as common corridors, linkways and public spaces, the perennial challenge is that a deep sense of community is something that comes about neither readily nor naturally.

In today's technology-enabled world, communities are no longer place-bound. With diverse modes of communication, efficient transport infrastructure, pervasiveness of the internet and social media, people form relationships and friendships in various settings—be it in the schools, workplaces and in the virtual or online world. The proliferation of sub-communities, in all form and manner, defies the planning wisdom and aspiration that propinquity would give rise to the formation of relationships, and that residential proximity would be a precursor to neighbourliness and social cohesion.

Nonetheless, in the pursuit of community resilience, the neighbourhood remains a necessary spatial construct for policy formulation and plan implementation. Without such a structural reference point, as a neighbourhood is, how do we logistically coordinate help for residents in times of crisis? How can we define the locale and community within which people can relate to each other and seek the services they need? How can we spatially organise the facilities needed for the recovery process? More importantly, how do we identify the weak, needy, poor, differently-able, vulnerable and aged without being able to determine where they belong?

Urban planners and architects have to continually seek and experiment with new built forms and design typologies to support relationship building, and would do this best within a defined physical realm. Likewise, in formulating urban policy, planning and design, as well as programmatic interventions, we have to harness spatial demarcations and defined proximities for effective execution.

Given this role in resilience building, neighbourhood planning has to go beyond the traditional curative approach of housing and amenities provision, to embrace a deeper understanding and awareness of issues such as racial segregation, gentrification, ghetto emergence, income inequality, differential access to opportunities, barriers to urban mobility and connectivity, definition of the urban commons, and participatory planning and decision-making.

In sum, a comprehensive approach that draws attention to the cross-sectoral challenges to neighbourhood resilience is needed, one that includes a consideration of community-based design of buildings and amenities, community-managed infrastructure such as common areas, parks and connectors, and wide ranging programmatic interventions in the diverse areas of education, sports, arts and culture, etc. Policies must be firmly grounded in robust research on the multi-faceted relationships and interactions of the physical environment with the people who live in them, and emplaced against a backdrop of prevailing social theories, sound planning models and thoughtful design practices.

3 Essays on the Ecology of Neighbourhood Resilience

The collection of multi-disciplinary essays in this book is the culmination of efforts by a group of researchers and professionals, along with their counterparts in government agencies in Singapore, as a response to the above discussion. Collectively, the essays highlight the key issues of and findings on community resilience based on recent research and projects. It serves to inform policy formulation and alert planning and design practitioners of the critical issues pertaining to community resiliency, in the context of the densely built housing estates of Singapore.

Community resilience, in this book, is discussed from three perspectives: (a) the *community perspective*, focusing on policies and practices that seek to forge social bonds, be it through grassroots activities, social services or racial integration; (b) the *place perspective* in terms of infrastructure provision in the areas of urban mobility, parks design and urban spaces for community interaction; and (c) the *planning perspective*, in terms of design interventions to enhance social encounters and respond to demographic changes, as well as resident engagement for stronger people-centric decision-making.

In the essays, these perspectives are embodied in four thematic areas, namely, community building, participatory planning, community ageing-in-place and designing resilient communities.

3.1 Community Building

As a young and dynamic society, Singapore continues to grow and change—environmentally, economically, socially and culturally. Concurrently, as each individual resident moves through the various life stages, the associated places of work, school, home and play, together with their meanings and inherent values, also change. While space displacement is often associated with the loss of place familiarity and erosion of social stability, relationships that are forged through non-place based activities and encounters can remain very enduring. In this regard, Tan and Teng's (2019) opening essay, *Fostering Social Cohesion in 21st Century Singapore*, which discusses the role of the People's Association (PA) of Singapore in strengthening social ties through a wide array of programs and activities, is highly relevant. The clear message is that participatory involvement in a variety of activities and programs can provide the milieu for relationship building and social bonding. These activities may range from leisure and recreational pursuits to educational, training and self-enrichment programs. It also provides the opportunity for strengthening people-government relationship through regular interactions and feedback, both formal and informal. But given that change in the city is inevitable and relentless, the paper highlights that policy makers need to be constantly cognisant of the new challenges that stem from an increasingly complex global environment, technological disruption, and changes in Singapore's social fabric and demographic profile. All of these can potentially create new social dynamics and undermine the efforts of building a resilient society based on trust and positive interactions amongst citizens, and between citizens and government.

In the context of Singapore's multi-racial population composition, another potential source of social tension is ethnic imbalance in the neighbourhoods of Singapore's heartlands. Leong et al. (2019) essay, The State of Ethnic Congregation in Singapore, discusses this with reference to the Ethnic Integration Policy (EIP) which seeks to manage racial clustering in order to maintain social harmony, a key element in resilience building. Using extensive GIS-based research which maps out present ethnic distribution across the various housing estates, the paper alerts readers to an anecdotally known but rarely publicly discussed fact that racial clustering continues to exist in various parts of Singapore. The studied correlation of such clustering against population age, income distribution and house prices is particularly telling, and lends credence to the concerns that ethnic clustering could perpetuate, and indeed, aggravate social cleavages. Potentially, this can be mitigated with directed policies to nudge ethnic redistribution by providing relevant amenities that are desired by particular targeted groups.

Would implementation of the right infrastructure to promote urban mobility, provide parks and urban spaces, and to support arts and cultural activities help to anchor population clusters or redirect their distribution? Three essays discuss this in the book.

Song et al. (2019) essay, *Urban Mobility and Resilience in Singapore: Transport Infrastructure Investment and the Demand for Travel*, based on their research on transport infrastructure and travel demand, highlights that enhanced mobility could be achieved through investment in transport infrastructure which in turn can promote greater community interaction and exchange. However, their research finds that for both work and non-work purposes, rail transit accessibility and expressway proximity have no considerable impact on household travel demand. The results suggest that improved accessibility brought by rail transit construction and expressway expansion may not stimulate or suppress the travel desire of local residents. While appropriate transport infrastructure can engender greater opportunities for exchange of goods and services, and movement of people and culture, the paper cautions that they do not in themselves determine travel demand, which is primarily influenced by households' socioeconomic and demographic characteristics. The research affirms that land use mix and population density have a positive influence, though small in magnitude, on household travel demand. Indeed, land use changes coupled with good urban design can promote connectivity, accessibility and walk ability, and thus directly or indirectly bring forth increased community interaction. In this regard, a relevant policy response is to support mixed land use strategies so as to provide diverse opportunities from a wide array of urban functions including employment, recreation and retail at the local level. With time, this promotes increased intra-town movement, which can engender convivial spaces and enhance social vibrancy. The role of informal and spontaneous exchanges in towns and cities cannot be over-emphasized; in transport planning, such exchanges can be promoted through investment in appropriate mobility-enhancing infrastructure, which may include non-motorised transport modes, such as cycling and walking.

The provision of urban green spaces is another aspect of infrastructure investment, and Sia et al. (2019) essay, *Social Resilience through Parks and Common Recreational Spaces in Singapore*, alludes to its role in resilience building. It highlights that in addition to providing ecological services, these open spaces offer a multi-sensorial environment that stimulates and improves community mental health. In particular, the multi-functional attributes of green spaces bring forth wide-ranging social benefits as they open up opportunities for users to interact with each other and to become attached to a place. Social interaction and place attachment are contributory factors toward social cohesion and a collective identity. A well-planned network of public green spaces also provides opportunities for encounters and interaction with nature, which is particularly important for people living in densely-built urban centres. The paper postulates that such people-nature interaction not only contributes to community and individual health, but can offer a 'protective factor' that helps city dwellers recover more quickly from adversity or stress.

Complementing the physical provision of urban spaces is the programmatic activities that are found in them. Here, Trivic's (2019) paper, *Bringing Arts Closer to Local Communities: Spatial Opportunities and Impacts on Community Bonding in Singapore*, is relevant as it discusses arts and culture in the city and its contribution to resilience planning. Using the conceptual lenses of place-making and cultural ecology, the essay discusses the roles and capacities of community arts and culture initiatives in supporting resilient local communities. Trivic alludes to place-making as a process of deliberate shaping of an environment to support social interaction, and to improve quality of life of local communities, empower residents through public engagement and boost civic pride. Place-making emerged in the 1960s as a response against the systematic destruction of people-friendly spaces; it has been used since the 1990s to describe a community-centric approach to public space design. The urban planning emphasis shifted from the physical attributes of spaces to their role in social and community life. In the research, the conceptual links between community arts and place-making is explored, with the goal to revealing and enhancing the underlying characteristics of the physical and social forms of local communities. The essay signifies 'creative place-making' as a process of bringing forth positive impacts to the community through deliberate and strategic creation of a symbiotic relationships between arts and culture, space and people. It notes that while creative place-making does not necessarily imply permanent changes to spaces, it always involves enhancing the sense of belonging and attachment to both neighbourhood and community.

3.2 Participatory Planning

Singapore today has a more engaged citizenry, which evidently appreciates a more pro-active role in shaping and serving society. Recognising this, Cho and Ho's (2019) essay, *Participatory Design to Co-create Community Spaces in Singapore*, presents a research project that aims to establish a participatory design approach to co-create community spaces in an existing neighbourhood. It is premised on the understanding that there is an inherent connection between participatory involvement in planning, decision-making and social resilience. The hypothesis is that residents, who have been actively involved in the community life of their immediate neighbourhood, are more likely to help others in times of need and would actively promote the well-being of their fellow citizens in the neighbourhood.

In this research, the participatory design approach was adopted by planners to co-create with residents two new community space typologies, namely, the neighbourhood incubator and social linkway. By specifically illustrating the processes and challenges in testing out new creative methods to engage the community, the essay offers insights into potential avenues for strengthening community's capacity to co-manage urban common spaces and to initiate change in the neighbourhood through collective action. The intensive engagement in the two real projects provided a deep understanding of what is meant by a 'sense of community'. Clearly, it is not just a social construct; it has a spatial dimension, as an amalgam of shared emotional contact through interaction with others in a particular place. It notably encompasses a sense of place attachment and a sense of membership whereby a community member feels and shares a 'right to belong'.

3.3 Community Ageing-in-Place

Demographically, Singapore is recognisably ageing. At both the individual and community levels, an ageing community has far-reaching impacts on social resilience. Taking a life course perspective of ageing, which considers the temporal evolvement of an elderly person over a lifetime, and not simply a snapshot account of old age, Fung's (2019) paper, *Place Familiarity and Community Ageing-with-Place in Singapore's Neighbourhoods*, postulates that community support for the ageing population is quintessential to the collective resilience of urban neighbourhoods. At the individual level, this community-oriented approach to ageing operates as a wellspring to draw upon against adversities; collectively, it is part and parcel of resilience building at the neighbourhood level. The approach to successful ageing would involve efforts to augment place familiarity as part of place-making to support shared community values and collective actions. *Fung* argues that by synthesizing time and space as a single continuum, place familiarity acts as a memory anchor for older adults, particularly in complex urban environments. More importantly, everyone in the community can share these experiences of place, albeit with different nuances, hence effectively developing a collective resilience of the urban commons. Thus, place familiarity becomes a factor contributing to such collective resilience, and this is illustrated in a tangible way with specific urban planning and design strategies, such as those for way-finding, perception-cognition of place, and physical interventions that reinforce familiarity, identity, memory and attachment. The absence of these design interventions is disaffection with place and social identity, reduced sense of safety and security, and diminished memory and social meaning.

3.4 Designing Resilient Communities

In the final essay, *Designing for Resilience in Public Housing in Singapore: An Architect's Perspective*, Siew and Nasution (2019) succinctly offer specific design experiences in public housing projects to draw out the physical manifestations of place familiarity and community-based thinking in site planning and neighbourhood design endeavours. Siew's reflections, as documented in this essay, are expressed through a personal account of childhood memories, design aspirations with their inherent value system, as well as practical architectural solutions. His honest belief that neighbourhoods are real and valuable urban constructs that can bring people together and have them form bonds with one another is inspirational for other designers. Architects and planners have to consciously design spaces to allow opportunities for people to meet and know each other, and it is in this process that they can help build community self-reliance and resiliency. When adversities strike, the community will come together and collectively overcome them. If this affirmative approach, as demonstrated in his way of embracing people's positive behaviour in the compact setting of a local neighbourhood, can be extrapolated to the city-wide scale, much more would be achievable.

While *Siew*'s reflections are centred on neighbourhood planning, and are peppered with vignettes of micro design elements such as common corridors, ground floor backyards, walkways and bus stops, he also highlights the importance of approaching design in an integrated system-based manner. To him, a key challenge in planning future resilient communities is to provoke a mindset change on the part of planners—i.e., not to think of housing, healthcare and eldercare as separate entities, but plan them as one ecosystem in order to maintain design integrity and social inclusiveness.

4 Planning the Future Urban Neighbourhood

As a whole, the book seeks to encapsulate the key constituents of resilient urban neighbourhoods in Singapore (and beyond). It is a dedicated effort to bridge research, empirical work, policy and practice, with some of the essays making references to scholarly models of resilience and governance philosophy.

Imagining future urban neighbourhoods with built-in resiliency requires the integration of policy, planning, urban and architectural design, programming, and community engagement at various levels. While social policies can activate and accelerate the socialization processes toward community development and resilience building, the physical processes of structuring, patterning and designing the future neighbourhood unit is innately challenging. Architectural and urban theories may abound on the complex relationships between people and their environment, but residents' response and behaviour, especially during times of crisis, is not readily predictable. The multi-faceted characteristics of resiliency planning demands deep knowledge of the intricate social interfaces and calls for empathetic design responses that can only come from experiential exposure to varied environmental conditions on the part of planners and architects. However, design professionals, on their own, can only do so much to foster creative place-making and promote place familiarity. Residents themselves, both collectively and individually, have a decidedly important role to play in resilience development. Through their routine interactions and daily activities in the ubiquitously planned neighbourhoods, they can consciously partake in meaningful interactions and, in the process, incrementally engender the sense of belongingness and community that would ultimately bring forth positive outcomes in times of adversity and crisis.

Acknowledgements The author would like to express her appreciation to Quyen Duong, Research Associate, Centre for Sustainable Asian Cities, NUS, for her assistance in compiling the academic references for this essay.

References

Australian Social Inclusion Board (2009) Building inclusive & resilient communities. Retrieved 9 March, 2017 from http://www.socialinclusion.gov.au/sites/www.socialinclusion.gov.au/files/publications/pdf/building-community-resilience-brochure.pdf

Carpenter B, Walker B, Anderies JM, Abel N (2001) From metaphor to measurement: resilience to what to what? J Ecosystems 4(8):765–781

Chelleri L, Olazabal M (eds) (2012) Multidisciplinary perspectives on urban resilience: BC3, Basque Centre for Climate Change

Cho IS, Ho KC (2019) Participatory design to co-create community spaces in Singapore. In: Leong CH, Malone-Lee LC (eds) Building resilient neighbourhoods: the convergence of policies, research and practice. Springer, Singapore

Folke C, Carpenter B, Walker B, Scheffer M, Chapin T, Rockstrom J (2010) Resilience thinking: Integrating resilience, adaptability and transformability. Ecol Soc 15(4)

Folke C, Carpenter S, Elmgvist T, Gunderson L, Holling CS, Walker B (2002) Resilience and sustainable development: building adaptive capacity in a world of transformations. Ambio 31(5):437–440

Fung JC (2019) Place familiarity and community ageing-with-place in neighbourhoods. In: Leong CH, Malone-Lee LC (eds) Building resilient neighbourhoods: the convergence of policies, research and practice. Springer, Singapore

IPCC (2012) Managing the risks of extreme events and disasters to advance climate change adaptation. In: Field CB, Barros V, Stocker TF, Qin D, Dokken DJ, Ebi KL, Mastrandrea MD, Mach KJ, Plattner G-K, Allen SK, Tignor M, Midgley PM (eds) A special report of working groups i and ii of the intergovernmental panel on climate change. Cambridge University Press, Cambridge, UK, New York, NY, USA

Kimhi S, Shamai M (2004) Community resilience and the impact of stress: adult response to Israel's withdrawal from Lebanon. J Community Psychology 32(4):439–451

Leong CH, Teng E, Ko WW (2019) The state of ethnic congregation in Singapore. In: Leong CH, Malone-Lee LC (eds) Building resilient neighbourhoods: the convergence of policies, research and practice. Springer, Singapore

Perry CA (1929) The neighborhood unit. Monograph One, vol 7. In: Regional survey of New York and its environs, neighborhood and community planning. New York, New York Regional Plan. Reprinted 1998, Routledge/Thoemmes, London

Resilience Alliance (2009) Assessing and Managing resilience in social-ecological systems: a practitioner's workbook, Version 1.0. Retrieved from http://wiki.resalliance.org/index.php/Main_Page
Sia A, Kua EH, Ho R (2019) Social resilience through parks and common recreational spaces'. In: Leong CH, Malone-Lee LC (eds) Building resilient neighbourhoods: the convergence of policies, research and practice. Springer, Singapore
Siew MK, Nasution IK (2019) Designing for Resilience in Public Housing in Singapore: An Architect's Perspective. In: Leong CH, Malone-Lee LC (eds) Building resilient neighbourhoods: the covergence of policies, research and practice. Springer, Singapore
Song S, Diao M, Feng CC (2019) Urban mobility and resilience in Singapore: Transport infrastructure investment and the demand for travel. In: Leong CH, Malone-Lee LC (eds) Building resilient neighbourhoods: the convergence of policies, research and practice. Springer, Singapore
Tan D, Teng E (2019) Fostering social cohesion in 21st century Singapore. In: Leong CH, Malone-Lee LC (eds) Building resilient neighbourhoods: the convergence of policies, research and practice. Springer, Singapore
The Community Social Planning Council of Greater Victoria (2013) Strengthening neighbourhood resilience: opportunities for communities and local governments. Retrieved 8 March, 2017 from https://resilientneighbourhoods.ca/wp-content/uploads/2017/03/Strengthening-Neighbourhood-Resilience.pdf
Timmerman P (1981) Vulnerability, resilience and the collapse of society: a review of models and possible climatic applications. Institute for Environmental Studies, University of Toronto, Toronto, Canada
Trivic Z (2019) Bringing arts closer to local communities: spatial opportunities and impacts on community bonding. In: Leong CH, Malone-Lee LC (eds) Building resilient neighbourhoods: the convergence of policies, research and practice. Springer, Singapore
UNISDR (2009) Global assessment report on disaster risk reduction. United Nations International Strategy for Disaster Reduction. Retrieved from https://www.preventionweb.net/english/hyogo/gar/report/index.php?id=9413
United Nations (2017) New Urban Agenda, presented at United Nations Conference on Housing and Sustainable Urban Development (Habitat III), Quito, Ecuador, 20 October 2016. Retrieved 23 November 2018 from http://habitat3.org/wp-content/uploads/NUA-English.pdf
Walker B, Abel N, Anderies JM, Ryan P (2009) Resilience, adaptability, and transformability in the goulburn-broken catchment Australia. Ecol Soc 14(1)

Fostering Social Cohesion in 21st Century Singapore

Desmond Tan and Eugene Teng

Abstract This chapter discusses how the People's Association (PA) has worked, and is working to foster social cohesion in Singapore. In particular, it focuses on the changes in contemporary Singapore that pose a challenge to the PA's work and how the organisation is now updating its approaches to adapt. In particular, the PA faces challenges from an increasingly complex global environment, technological disruption and changes in Singapore's social fabric. To keep relevant, the PA is shifting towards deeper relationship building and strengthening its outreach to more segments of society. It is also concurrently upgrading its Community Clubs located across the country while also taking its programmes beyond the clubs and into the neighbourhoods. By building social cohesion and a strong grassroots movement, the PA contributes to the overall strengthening of resilience among Singaporeans.

Keywords Cohesion · Social resilience · Grassroots · Community outreach

1 Introduction

Community resilience is commonly defined as the collective ability to cope with, or overcome adversity. Many factors are able to mediate the negative influence of hardship, including on individual's psychological make-up. However, within the context of a lived-in neighbourhood, social cohesion—the strength of social bonds between members of the group—undoubtedly plays a significant role. A cohesive

---

Eugene Teng collaborated in the writing of this chapter by transcribing Desmond Tan's perspectives and reflections, and by including relevant academic references.

D. Tan (✉)
People's Association, Singapore, Singapore
e-mail: Desmond_TAN@pa.gov.sg

E. Teng
Institute of Policy Studies, National University of Singapore, Singapore, Singapore
e-mail: eugeneteng.yl@gmail.com

C.-H. Leong and L.-C. Malone-Lee (eds.), *Building Resilient Neighbourhoods in Singapore*, Advances in 21st Century Human Settlements,
https://doi.org/10.1007/978-981-13-7048-9_2

society is one that works towards the well-being of all its members, creates a sense of belonging, promotes trust and fights against the marginalisation within and between different groups (OECD 2011). Such a society would encourage cooperation and tolerance between its members, and make them more willing to help one another in times of crisis. Studies have shown social cohesion within a community to be strongly associated with resilience (e.g. Ahmed et al. 2004; Townshend et al. 2015; Patel and Gleason 2018).

Given the importance of social cohesion in building a resilient nation, the People's Association (PA) was established in Singapore in 1960, shortly after the country attained full internal self-governance in 1959. At the time, Singapore was divided along racial and religious lines. Many people lived in ethnic enclaves, which largely determined who they mixed with, the schools they attended, and the jobs they held. The objectives of the PA were thus to bring the Government and its activities to the people, with the aim of keeping in constant touch with the people to know their grievances, and to inculcate social qualities in building up a cohesive society. Through its network of grassroots leaders and community partners, provided platforms for social bonding among the different ethnicities through participation in group activities at the local and national levels, to foster community bonding and strengthen social cohesion (Singapore Statutes 1960).

This chapter discusses how the PA builds social resilience by fostering and maintaining social cohesion in Singapore. Section 2 gives an overview of the history of the PA; Sect. 3 identifies the modern-day challenges that PA has to deal with and Sect. 4 explains how the PA is currently addressing those challenges. Finally, Sect. 5 concludes by highlighting the importance of grassroots mobilisation in fostering social cohesion.

2 Brief History of the People's Association

In the beginning, the PA took over 28 community centres (CCs) that were originally built by the British colonial government as recreational spaces. The PA developed them to serve as a common space for social mixing and as a hub for recreational activities. CCs offered practical skills courses, such as dressmaking and radio repair, and essential services like medical vaccinations, clerical help, and telephone lines. The PA also provided the government a means with which to stay in constant touch with the people and their grievances, with CCs serving as a contact point between government leaders on one hand, and grassroots leaders and the people on the other. Over the years, the PA would rapidly expand the number of CCs around the island (Figs. 1 and 2).

PA's mission to maintain social cohesion would come into the fore in 1964, when race riots broke out. Besides the fighting in some areas, Chinese and Malays in other regions of Singapore began to see each other with distrust and fear. PA played an important role in mitigating the situation by forming Goodwill Committees in each constituency. These committees consisted of community leaders of all ethnicities, who went around their neighbourhoods calming tensions by dispelling rumours,

Fig. 1 Early days of the Community Centre where people gathered to watch television. *Source* People's Association

Fig. 2 Dressmaking courses offered at CCs as more women entered the workforce. *Source* People's Association

distributing rations and assuring their residents of the situation and their safety. CCs served as a crucial safe and neutral space for residents during the riots, where volunteers could also meet to strategise ways to reduce racial tensions. After the riots, the Goodwill Committees were formalised into the Citizens' Consultative Committees (CCCs).

Over the following decades, as Singapore modernised, the PA's role grew in scale and complexity. Beyond the CCCs, more grassroots organisations were created to take on various specialised roles; one such grassroots organisation is the Community Emergency and Engagement Committees to raise emergency preparedness in each neighbourhood. Community Centres were upgraded and rebranded as Community Clubs, each housing a diverse range of facilities like multi-purpose halls, conference rooms, dance studios, and fitness rooms. Courses offered at the CCs also kept up with the times to suit the changing needs and interests of the residents.

The effectiveness of the PA's work was clearly visible during two public health crisis—the SARS epidemic and the haze outbreak. The outbreak of SARS (Severe Acute Respiratory Syndrome) in 2003 infected 238 people and claimed 33 lives. Despite the potential health dangers to themselves, around 650 volunteers from the grassroots organisations conducted daily temperature checks on 22,000 people for 10 days. They also carried out contact tracing, distributed food rations to those quarantined at home, and organised dialogues to allay residents' concerns. Because grassroots leaders were familiar with the residents within the constituency, they were able to help the government distribute resources promptly to the targeted groups.

Another example was in June 2013, when regional haze caused record-high levels of pollution in Singapore. The government specially prepared protective masks to be given out to low-income families, and the PA, through its grassroots network, distributed such masks to these families within a very short time. The grassroots advisers and grassroots leaders also taught these residents how to use the masks correctly during the house visits, and made sure they were able to gain access to key necessities by distributing a WeCare pack comprising rations and face masks. In both of these incidents, the speed and willingness with which PA mobilised the volunteers, exhibited the community's resilience in the face of crisis. Such selfless action is unlikely to occur if the community was not socially cohesive (Fig. 3).

The PA has a dual funding mechanism. Grassroots organisations, which are managed by volunteers, generate income through fund-raising activities and collection of fees for courses conducted at CCs. This self-funding aspect ensures community ownership of grassroots initiatives, as these funds mainly go into organising community programmes and activities to cater to the needs and interests of residents. The Singapore Government supports community initiatives by co-funding programmes organised by grassroots organisations as well as the cost of building and upgrading community infrastructure. The government's funding represents its commitment to promoting social cohesion among Singaporeans.

Today, the PA's mission of building and bridging communities in Singapore remains unchanged. At the same time, it has been evolving its methods with the times, to stay relevant and effective amidst new social trends and challenges.

Fig. 3 Volunteers go door-to-door to distribute WeCare packs containing rations and N95 during the SARS outbreak. *Source* People's Association

3 Contemporary Challenges to Cohesion

3.1 More Complex Global Environment

The geographical and demographic realities of Singapore put her in a unique position in today's highly interconnected world. Globally, we are facing greater and more complex uncertainties—terror attacks, resurgence of populist politics, and tensions between superpowers. Singapore is not immune to these threats. Our open borders mean that international and regional geopolitical developments have the potential to affect social cohesion. Achieving PA's purpose and mission to bond and bridge communities amidst such an operating and complex global environment will only become more challenging in the future.

Fundamentally, Singapore is a small nation-state surrounded by larger countries in the region, and given the lack of a hinterland or natural resources, our economy is also subject to the vagaries of global trade. When the economy is good, there can be greater social mobility, economic opportunity, and real income growth, all of which leads to more overall happiness, optimism and trust, which can in turn strengthen social cohesion (Abrams and Vasiljevic 2014). On the flipside, if the economy declines, social cohesion weakens.

Additionally, given Singapore's multi-racial and multi-religious societal makeup, we also have to be aware of global trends that may divide the population.

This includes appeals by other nations to local ethnic communities within Singapore, as well as religious sects that might promote exclusivist or radical ideologies. For example, in recent years, there has been a rise in radicalised Singaporeans arrested (Loh 2017), and the PA sees its role as to help ensure that such incidents do not provoke distrust of certain racial or religious communities within Singapore. It will try to help Singaporeans understand what is in Singapore's best interests and to hold communities together even as international developments may cause some to drift apart.

3.2 Technological Disruption

The pace of technological change has transformed how Singaporeans consume information and interact with one another. Digital communities are forming alongside physical ones. On one hand, this creates more opportunities to bring people together, but on the other hand, people tend to only talk to those who share similar views, leading to echo chambers. This increases polarisation as online interactions tend to reinforce existing views without sufficient moderation from alternative opinions. Additionally, given the pervasive use of social media to disseminate and consume information, "fake news" has tremendous potential to destabilise communities within a short span of time. These developments mean that PA has to derive new ways to engage residents beyond face-to-face engagement, and that it needs to develop online networks to communicate reliable information in an age of misinformation.

3.3 Changing Texture of Society

The texture of Singaporean society is evolving in many ways. First, our demographic makeup is changing. We face an ageing population, with the old-age support ratio[1] likely to change rapidly from 4.2 in 2018 to 2.0 in 2040. This means that there is a greater role for PA to play in keeping seniors socially engaged and physically active.

Second, whereas in the past social integration was focused on the different racial groups, there is now an additional dimension of integrating new immigrants with native-born citizens. This is given the current pace of immigration as well as the rise in transnational marriages among Singaporeans. Integrating new citizens within the Singaporean social fabric can be challenging. As we import goods and services, we also import the culture and way of life of immigrants. Some people may feel uncomfortable about the different cultural norms and social habits that new

[1] The old-age support ratio refers to the number of people who are capable of providing economic support to the number of older people who may be dependent on others' support. In the current Singapore context, it is the ratio of the working age (20–64 years-old) population to population aged 65 years and older. A lower old-age support ratio means a higher dependency on each working age person per old-age person.

immigrants and foreigners may bring with them, while new immigrants may take time to familiarise with and adapt to Singapore's norms and values. A 2010 survey found that significantly fewer naturalised citizens than local-born citizens believed that speaking English, completing National Service,[2] and neighbourliness were important to being Singaporean (Leong and Yang 2015).

Third, new social cleavages may be emerging. In addition to the traditional cleavages of race and religion, a recent study by the Institute of Policy Studies (2017) found that most Singaporeans' social networks are bounded by their housing type and educational level. Along with ongoing efforts to keep income inequality in check, all these point to a potential social divide by the economic (income) class of the people. PA's role therefore is to encourage social mixing by maximising the common space for residents of all backgrounds to come together, as well as to mobilise residents in each community to help the less fortunate so as to co-create a caring and compassionate society.

Lastly, we now have a more engaged citizenry in Singapore and a proliferation of social causes. Compared to the 1960s and 1970s, Singapore's citizens today are more educated, and are increasingly making their voices heard and desiring a more proactive role in shaping their society. Among the youth (defined by national average as those aged 35 years and below) especially, there has been an uptick in civic engagement, such as contacting a government official, writing to a newspaper or magazine, or commenting on an online forum or blog, since 2005 (National Youth Council 2014). Social media also lowers barriers to civic participation and gives voice to even the most obscure viewpoints. Consequently, there are more avenues to speak up, and more groups are making use of them, lending to more vibrant public discussions and more issues being aired. There are now interest groups that advocate for animal welfare, heritage preservation and foreign worker rights, which did not exist in the past. The challenge for PA therefore is to help create a safe space for views to be heard and aired in a respectful manner, while keeping society cohesive.

4 Evolving with the People

As the Singapore's society evolves, the PA has to evolve along with it in order to stay relevant to the people and community. What has not and will not change are the core guiding principles of promoting people-to-people bonds and being a two-way bridge between the government and the people. The precise approach, however, must continuously be updated. Since 2015, the PA's main strategy has been to find ways to go beyond: beyond events, beyond "regulars", and beyond CCs.

[2]Mandatory two-year conscription for all male Singapore citizens and permanent residents at the age of 18, into the military, police or civil defence services.

4.1 Beyond Events—Building Relationships

The PA's grassroots organisations are well-known amongst Singaporeans for the many events they organise for residents to get together and mingle. These range from large-scale dinners for festive occasions like Chinese New Year and Singapore's National Day, to cosier gatherings over block parties, neighbourhood barbeques, resident excursions, or even the football World Cup and movie screenings. Such initiatives provide a platform for residents of all backgrounds to mingle and interact, and are also an opportunity for grassroots leaders to engage with residents on their needs, concerns and aspirations. In these ways, organising events is important not as an end in itself, but as a means toward building and strengthening relationships between residents and the community.

To achieve its overarching goals, the PA is consistently seeking to improve the design and conduct of programmes. It is acknowledged that the success of programmes and activities organised should not be measured by size of attendance, but the quality of relationships built. This approach is exemplified in the current way in which Edusave[3] award ceremonies are conducted. In the past, each ceremony hosted over a thousand families for three to four hours, with much of the time spent watching children go up on stage one-by-one to receive their awards. Now, award ceremonies are conducted more often, with each one hosting smaller groups of 30–50 families. Beyond just distributing the awards, these gatherings feature sharing sessions where parents talk about the experience of parenting their child, and recipients show their appreciation to their parents. Such events are more intimate and engaging, and allow for greater interaction amongst families (Fig. 4).

Another way of building people-to-people relationships is through getting residents to take greater ownership of the community in which they live, with volunteers playing the role of facilitating the identification of community issues and leading the mobilisation of residents to take action. One example is the case of a cluster of seven condominiums in West Coast facing a road widening project. With an original hypothesis that residents would be most concerned about disamenities during the construction process, the local Neighbourhood Committee decided to deepen engagement of the condominium residents to better understand their concerns. They discovered that the real concern was the loss of 100 trees which were an important characteristic of the neighbourhood. With that in mind, the grassroots volunteers worked with the residents and government agencies to co-create a solution. They planted 700 trees to green the estate, more than making up the loss of the original 100. When the road widening project was completed, they held a mass tree planting ceremony involving 1500 residents, all of whom took ownership and contributed to beautifying their local neighbourhood. Subsequently, the residents from the various condominiums started exploring more ways of collaboration, such as coordinating their mosquito fogging. As a result, the entire community benefitted (Fig. 5).

[3]The Edusave Scheme was introduced by the Government of Singapore in 1993 to provide every student with financial resources to participate in school enrichment activities. It also annually gives out merit awards and bursaries to exemplary students.

Fig. 4 Parent-child bonding at the Edusave Award Ceremonies. *Source* People's Association

Fig. 5 Neighbourhood Committee in West Coast leading efforts to plant 700 trees after road widening project. *Source* People's Association

Aside from intra-community bonding, the PA is also responsible for being a conduit between the people and the government. This entails having grassroots leaders communicating government policies to the ground, and gathering and conveying residents' feedback on policies to the government. In this regard, it is thus important to strengthen relationships between grassroots leaders and the people so that the people trust these volunteers to be their voice and can rely on them to access relevant policy information. The PA is placing greater emphasis on this aspect of community engagement. Grassroots leaders have been proactively reaching out and engaging residents in a more personal way. This means frequently going on house visits, and having small group dialogues and conversations in informal and casual settings, such as at hawker centres and coffee shops. PA also organises regular "Kopi Talks", which are topical dialogue sessions chaired by a Minister, for grassroots volunteers to raise queries in order to understand policies better so that they can in turn better explain them to their residents and address their concerns. Local communications workgroups within each constituency have also been set up, consisting of volunteers specially tasked to spearhead policy communication efforts and better understand residents' needs, concerns and aspirations. These grassroots volunteers are better trained and equipped to engage residents on policy matters. To complement the face-to-face outreach by grassroots volunteers, the PA also produces handy guidebooks on relevant policies that residents can refer to. These "Ask Kopi Kaki" booklets are organised by life-stage rather than policy domain—one for seniors, one for young families, and one for working adults.

4.2 Beyond "Regulars"—Broadening Outreach

Over the years, CCs and PA activities have been well-patronised by "regulars"—generally middle-aged to elderly residents living in the nearby public housing estates. Beyond these groups, the PA aims to be more effective in engaging others who are less involved in PA activities, such as Professionals, Managers, Executives and Technicians (PMETs), young families and ethnic minorities. A good illustration would be the Embracing Parenthood Movement (EPM), which gives PA a good platform to engage young families with children who are around the same age. These residents may live in close proximity to one another and yet do not know one another well. Through this support network, parents can potentially turn to one another for advice and support rather than simply relying on online groups. Hence the EPM creates a starting point for the formation of local smaller communities through their parenting journey.

Many residents also spend their time getting involved with organisations and associations that are not a part of the PA. This is a welcome development as they are forming social bonds and in many cases also contributing to society. For example, some Muslims and Christians are highly active in charitable activities organised by their mosques and churches respectively. Others volunteer with cause-based groups, such as those concerned with migrant worker rights and environmental awareness.

Still others involve themselves with hobby or interest groups, such as hiking, board-gaming, or bird-watching. PA's intent, therefore, is to explore ways to collaborate with the wide variety of other social and community organisations across the island. In the case of hobby and interest groups, the PA tries to facilitate them using their interests to further a social cause, or to involve residential communities in some way (e.g. cycling for charity, organising free classes). For those groups already involved in social causes, PA grassroots organisations will try to coordinate with them better so as not to duplicate programmes. The PA can also partner with these organisations to conduct joint activities. The objective of working with all these organisations is to ground some of their activities to geographical neighbourhoods, so that socialisation can also occur between neighbours.

4.3 Beyond CCs—Going to the People

A major pillar of the PA's operations is the CCs. From the beginning, CCs were created so that every Singaporean could gain access to recreational facilities or take part in social activities at affordable rates. Therefore, CCs have always served as gathering points for the community, and thus a platform for grassroots leaders to get to know residents and their needs better as well as to disseminate government information.

With growing aspirations and Singaporeans becoming more sophisticated in their lifestyles, CCs have also evolved with their offerings to stay relevant. One way the PA is updating CCs is to increasingly co-locate CCs with libraries, police centres or even to primary schools to provide residents easy access to a range of amenities in a single location. At the town level, PA has begun setting up "integrated lifestyle hubs", and has designed the CC as a central part of the town centre's facilities along with partners like Sports Singapore, Ministry of Health, National Library Board and private vendors to meet sports and recreational needs, whilst leveraging the higher human traffic at transport nodes. In 2017, the first of these "Hub CCs" was opened in Tampines. Named 'Our Tampines Hub', the sprawling complex houses a stadium, multiple sports halls, a library, a mall, a clinic, a public service centre holding seven key agencies, and much more. Having so many different amenities in the CC makes them more attractive and relevant to a larger number of people.

In addition to Hub CCs, the PA is also setting up specialist CCs to appeal directly to specific types of residents. For example, the PA will soon complete imPAct@Hong Lim Green, which is targeted at PMETs working in the Central Business District. imPAct@Hong Lim Green is intended to provide a convenient after-work destination for rest and recreation. Its central feature is a CrossFit gym, and it will also house an alfresco café, workstations, and a fitness studio for programmes like aerial yoga. In the future, other specialised CCs will be developed to target other population profiles, such as young families. These CCs would complement existing Hub CCs to provide a variety of options for all segments of the population. Like Hub CCs, specialised CCs will help attract new residents to CC activities (Fig. 6).

Fig. 6 Evolution of CCs (from top to bottom): Tuas CC in 1960 (*Source* Kampong Days), Our Tampines Hub in 2017 and PassionWave specialist CC for water sports. *Source* People's Association

However, in the midst of updating and improving the CCs, the PA and the grassroots movement acknowledge that they must not fall into the trap of becoming overly dependent on the CC as a platform for outreach and activities. In the past, CCs used to be the lifestyle highlight of the town—partly because of the programmes and amenities offered, but also because there were not many alternative options for leisure and recreation. These days, most town centres have multiple malls, and there are many other organisations and associations offering courses and recreational facilities. Furthermore, better transport infrastructure has improved residents' mobility. No longer are they confined to the one CC in their town centre. With all the choices Singaporeans now have, it can be difficult to draw people to the CC sometimes. The CCs cannot be everywhere, and the PA must move away from the fixation that everything must be organised at the CC, lest residents think that the PA can only be reached at the CC and nowhere else. People already gather for different purposes at different places, so the challenge is how the PA can bring its outreach and activities to places that are more convenient for them.

Even as the PA modernises its CCs and programme offerings, it is also expanding the range of community spaces in which it operates. In short, it is making its presence felt where residents organically congregate. If residents are not coming to the CCs, then the PA will go to where they are. Activities and events are organised at common spaces closer to residents, such as neighbourhood amphitheatres, coffee shops, playgrounds and void decks.[4] Exemplifying this approach, a group of grassroots volunteers in Zhenghua took to setting up a stand at bus stops serving free coffee to residents in the morning peak hour. This allows the volunteers to engage residents' for feedback on local municipal issues, and also build rapport which may otherwise not be possible due to time constraints to participate in grassroots activities. In this way, people can feel that the PA is present in their lives and listening to them, without having to physically go to a CC.

5 Building Resilience at the Grassroots

As the world becomes more complex and Singaporeans' lifestyles become more sophisticated, the PA needs to update its strategies to stay relevant to the people so that it can carry out its work of fostering social cohesion. In totality, going beyond events, "regulars", and CCs can help the PA to broaden its reach and deepen its connection with the residents. Broadening outreach will help the PA to keep up with the increasing diversity in society. Deepening connections will improve and reinforce the strength of interpersonal and intergroup bonds, be it between people or between people and the government. With greater and more extensive social trust and rootedness to their communities, social cohesion will be strengthened.

[4]Void decks are ubiquitous to Singapore's public housing estates. They are open, communal spaces on the ground floor of every housing block.

Being able to trust one's neighbours and feel a sense of belonging to Singapore and the immediate community is important for building resilience at the grassroots. Trust enables people to demonstrate civic sense even when they disagree and provides a ballast against influences that threaten to sow discord amongst people. Similarly, rootedness and sense of belonging creates a shared identity among otherwise disparate peoples. It also helps residents feel they have a stake in their community and be more willing to contribute to enhancing its wellbeing. The PA's contribution to social resilience framework in Singapore is thus to foster and maintain social cohesion among the people.

Integral to the PA's role in strengthening resilience is the extensive grassroots network that has been built over the decades. Having a strong grassroots movement is essential to resilience because it means having a ready network that can be mobilised to assist the community in times of crisis. The effectiveness of the PA's grassroots mobilisation has been clearly seen in past crises, such as the Hotel New World collapse (Lim 2011) and the SARS outbreak. Equally important is what the grassroots movement reflects—a strong sense of commitment to the community. Grassroots leaders and volunteers are themselves residents, and they want to make a difference to their communities. That these residents are willing to volunteer their time is a testament to the sense of belonging they have toward their neighbourhoods and the high degree of trust in their neighbours. It is thus important to support the grassroots movement so that they inspire others to do the same, and consequently strengthen social cohesion.

Internationally, Singapore was an active participant in the preparation of the United Nations' New Urban Agenda in 2016 (United Nations 2017). Many government arms are involved in working towards its various goals, such as sustainable urban planning and tackling greenhouse gas emissions. The PA plays a part in this collective effort by focusing on strengthening social resilience to mitigate against natural and man-made disasters. The grassroots-led Community Emergency and Engagement (C2E) Committees actively plan disaster responses and organise public exercises to prepare residents for emergencies. When disasters strike, the C2E Committees will be the first on the ground to coordinate response efforts and assist the authorities in defusing tension and restoring public confidence. Through improving local responses to disasters, the PA contributes to local and international sustainable development goals.

In summary, the 21st century brings with it many challenges that could potentially weaken the social bonds formed between neighbours. The PA therefore remains focused on its core mission of social cohesion and racial harmony, while ensuring its approaches are in keeping with the latest developments in society. By going beyond events, beyond "regulars", and beyond CCs, the PA is widening its outreach, growing its capabilities, and deepening its relationship with residents, as it works towards building a cohesive society where everyone regardless of background, view themselves and others as 'One People, One Singapore', a shared identity and common destiny that binds us all.

References

Abrams D, Vasiljevic M (2014) How does macroeconomic change affect social identity (and vice versa?): insights from the European context. Analyses Social Issues Public Policy 14(1):311–338

Ahmed R, Seedat M, Van Niekerk A, Bulbulia S (2004) Discerning community resilience in disadvantaged communities in the context of violence and injury prevention. South African J Psychology 34(3):386–408

Institute of Policy Studies (2017) A study on social capital in Singapore. Institute of Policy Studies, Singapore

Leong CH, Yang WW (2015) Social markers of integration: what matters most to Singaporeans? In: Yap MT, Koh G, Soon D (eds) Migration and integration in Singapore. Routledge, Oxford, pp 39–64

Lim TS (2011) Hotel New World collapse. Singapore Infopedia. Retrieved from: http://eresources.nlb.gov.sg/infopedia/articles/SIP_783_2009-01-02.html

Loh D (2017, September 13) Individuals in Singapore being radicalised at faster pace: Shanmugam. Channel NewsAsia. Retrieved from: https://www.channelnewsasia.com/news/singapore/individuals-in-singapore-being-radicalised-at-faster-pace-9213840

National Youth Council (2014) The State of Youth in Singapore 2014. National Youth Council, Singapore

OECD (2011) Perspectives on global development 2012: social cohesion in a shifting world. OECD Publishing, Paris. Retrieved from: http://dx.doi.org/10.1787/persp_glob_dev-2012-en

Patel RB, Gleason KM (2018) The association between social cohesion and community resilience in two urban slums of Port au Prince, Haiti. Int J Disaster Risk Reduction 27:161–167

Singapore Statutes (1960) People's Association Act. Retrieved from: https://sso.agc.gov.sg/Act/PAA1960

Townshend I, Awosoga O, Kulig J, Fan H (2015) Social cohesion and resilience across communities that have experienced a disaster. Nat Hazards 76(2):913–938

United Nations (2017) The New Urban Agenda, A/RES/71/256. Habitat III and United Nations

The State of Ethnic Congregation in Singapore Today

Chan-Hoong Leong, Eugene Teng and William Weiliang Ko

Abstract Social harmony has been the bedrock of Singapore's prosperity and success since the city-state gained independence in 1965. Being a multi-racial and multi-religious society, there is a political impetus to ensure that Singaporeans have opportunities to interact with people from other racial or socio-economic background. This is crucial in order to promote cultural empathy and understanding, and consequently greater social resilience. For the policymakers, a key platform to achieving this objective is via its residential policy, with the aim to avert the formation of ethnic enclaves, which are known to be common fault lines of social tensions. The Ethnic Integration Policy (EIP) was implemented in 1989 to ensure each block of public housing has a mixture of households from all races. While the EIP has served its intended purpose, the analyses in this paper reveal that some neighbourhoods continue to show a higher concentration of selected ethnic communities, which may, in the long-run, impact social cohesion. The paper offers a few suggestions on urban planning and intercultural engagement.

Keywords Ethnicity · Integration · Intergroup contact · Diversity · Multiculturalism · Residential · Congregation

1 Introduction

Singapore is a diverse society by any socio-metric indicator. The city-state and former British colony is home to some 3.9 million residents, the majority of which are of

C.-H. Leong (✉) · E. Teng
Institute of Policy Studies, National University of Singapore, Singapore, Singapore
e-mail: leong.chanhoong@nus.edu.sg

E. Teng
e-mail: eugeneteng.yl@gmail.com

W. W. Ko
Department of Geography, National University of Singapore, Singapore, Singapore
e-mail: weiliang.ko@gmail.com

C.-H. Leong and L.-C. Malone-Lee (eds.), *Building Resilient Neighbourhoods in Singapore*, Advances in 21st Century Human Settlements,
https://doi.org/10.1007/978-981-13-7048-9_3

Chinese descent (74%), followed by a sizeable proportion of ethnic Malay (14%), Indian (9%), and other minority races (3%) (Department of Statistics Singapore 2017). The multi-racial landscape is an inheritance and historical legacy of Singapore since its founding years.

Before Singapore became a self-governing state in 1959, the British colonial government segregated the residential locations of the major races (Chinese, Malay, and Indian) in distinct parts of the island. The Chinese were concentrated around the downtown area, Malays in Kampong Glam and Geylang Serai, and the Indians in the Serangoon and Sembawang districts. The racial zonal distribution was a part of the Britain's "divide and rule" strategy—it created a dependency on the colonial government for all ethnic related affairs and it helped consolidate the colonial influence and power over all matters on law and order. Not surprisingly, the spatial segregation was a fertile ground for inciting racial conflicts and misunderstanding. Ethnic enclaves were believed to be the key factor behind the tumultuous racial riots in the 1950s and 1960s (Mutalib 2004; Ooi 1993; Van Grunsven 2000).

After Singapore gained independence in 1965, the supply of public houses was ramped up significantly through the Housing Development Board (HDB). High-rise residential apartments were built across the city-state, including previously uninhabited land in the country. As of 2010, 87.2% of all residential households in Singapore are homeowners, and approximately 80% of them reside in public housing, or HDB flats (Department of Statistics Singapore 2015). Newly-built apartments are offered to all Singaporean citizens at an affordable price, supported by a suite of mortgage options and special housing grants and subsidies. The resale of HDB apartments were initially controlled but gradually liberalised over time. In 1979, HDB owners could sell their flats in the open market after completing a 5 year minimum occupation period (Phang and Kim 2013). Residents with the means and improved socio-economic conditions could purchase a larger, more expensive, HDB apartment.

2 Housing Policy and Racial Harmony

The objective of public housing—and by extension, that of home ownership—goes beyond providing an affordable roof over one's head. It was designed to introduce multi-racial living among residential neighbourhoods and nurture a sense of belongingness. It makes Singaporeans stakeholders in the city-state, and offers them a cause to feel rooted (Yuen 2007). More crucially, HDB housing is also an important vehicle to promote a harmonious and cohesive living environment. It follows the view that a spatial mixture of residents from diverse racial backgrounds will lead to more intergroup contact which would then result in greater awareness for other cultures, religions and norms (Mutalib 2004; Ooi 1993). With enhanced interaction and integration, a communal identity can also be forged (Siddique 1993). Today, the large percentage of public homeowners is a testament of this success. The HDB dwelling is a quintessential component of the Singaporean identity, and it cemented its role as a pillar of social harmony, a sacred social asset in Singapore.

Table 1 Ethnic quota, population distribution, and percentage residing in HDB dwellings

Races	Block limit (%)[a]	Neighbourhood limit (%)[a]	Population distribution (%)[b]	Reside in HDB swelling (%)[c]
Chinese	87	84	74	81.3
Malays	25	22	14	96.8
Indians/others	15	12	12	82.7

Note
[a]EIP limit is based on the revised quota in 2010
[b]Includes Singaporean citizens and permanent residents
[c]Percentage of HDB dwellers for each races obtained from 2010 population census; HDB dwellings include apartments of all sizes, i.e., 2–5 rooms, HDB Executive Apartments/Mansionette
Source Singapore Parliamentary Debates (2010)

However, by the mid-1980s, there were signs of emerging ethnic enclaves as selected districts became more attractive to residents of certain racial groups. In 1989, the then-Minister for National Development produced statistics showing stark ethnic preferences in the resale market; Chinese flat buyers were overrepresented in central areas such as Hougang, Bukit Merah, and Redhill, Malay flat buyers were overrepresented in peripheral areas such as Bedok, Tampines, and Taman Jurong, and Indian flat owners were overrepresented in Yishun and Kampong Java (Sin 2002a). The list is not exhaustive and the proclivity to live among people from the same ethnic background also emerged in other neighbourhoods. Based on these trends, the HDB forecasted that by the year 2000, Bedok would become a Malay "enclave" with 43% of the population Malay. Similarly, Bukit Merah would wind up with a 93% Chinese population, and Indians/Others would be overrepresented in Kampong Java at 29.8% (Ooi 1993).

In an attempt to nip racial congregation in the bud, the Ethnic Integration Policy (EIP) was introduced in 1989. The policy requires the ethnic composition for each block of HDB flat and within the proximal neighbourhood to follow the national ethnic distribution (i.e., Chinese, Malay, Indian and Others) as closely as possible (see Table 1, on ethnic ratio in each block and neighbourhood). From Table 1, each block of flats shall not have more than 87, 25, and 15% of households that are of Chinese, Malay, and Indian/Other racial ancestry, respectively, and the neighbourhood level, 84, 22, and 12%, for the three races. The ratios are interlocked as having one more ethnic group means less for the other. As such, there is a buffer for each ethnic quota (e.g., 74% population is Chinese, but 84% block limit for Chinese), and the total percentage at block and neighbourhood levels for all ethnic groups exceeds 100%. Total block and neighbourhood limits for all races are 127% (Chinese 87%, Malay 25%, Indian/Others 15%) and 118% (Chinese 84%, Malay 22%, Indian/Others 12%), respectively. Home owners may also appeal to the housing authority for a waiver of EIP if there is a compelling reason for an exception (Mohan 2018).

How does the EIP work? When an apartment block reaches the maximum quota for a particular racial category, a prospective buyer of that ethnic identity will not be allowed to purchase a resale flat from an existing home owner of a different racial category, although the prospective buyer could still acquire a unit from a seller of the same ethnic group. For instance, if a block has reached the maximum quota for ethnic Chinese, a prospective Chinese buyer would not be allowed to purchase a resale flat from a Malay household as this would further dilute the minority representation in the block, but the buyer could procure a unit from a Chinese seller, as it would have no impact on the existing ethnic balance.

The policy seeks to ensure that no single-race block will ever exist, and there are social and spatial opportunities for residents to interact with people of other races in their daily lives (Yuen 2007). This policy supposedly mitigate the spatial concentration of races among the HDB public housing estates and prevent further deterioration of ethnic over-representation among affected neighbourhoods. Each block is intended to be a small microcosm of Singapore's society at large, and the quota is in place to reinforce the goals of the HDB housing policy in fostering greater racial acceptance and tolerance (Lai 1995; Ooi 1993; Siddique 1993; Sim et al. 2003).

It should be added that the EIP is targeted at the resale market, whereas the sale of the newly constructed flat is under the direct purview of the housing authority. Operating in a similar fashion as the EIP, the latter calibrates the racial quota of homeowners in each new residential building but there are more options to control and allocate (e.g., through balloting, as opposed to market forces) apartments to suit a more narrow range of new home-buyers (e.g., priorities for young families and first-time buyers). The resale market on the other hand is more susceptible to the rogue market forces, and thus more sensitive to price mechanisms.

3 Taking Stock of EIP and Its Collateral Effects

Has the EIP mitigated racial clustering? Almost 30 years on, there is limited empirical data or systematic study on the spatial distribution of ethnic categories across Singapore. However, based on the demographic census, it is a fact that there are disproportionately more Chinese living in central parts of Singapore, and Malays in the north and the east, for instance. Anecdotally, locals are also aware of areas that are more "Malay" or "Indian", largely corresponding with the 1980s trends that prompted the policy in the first place.

Sin (2002b, 2003) examined residents in Bedok New Town using data from the 1980, 1990, and 1996 registers of electors. Collecting individualised data in this fashion is appropriate in the Singapore context as voting is compulsory and the registers are computer-generated from central government databases. However, these names did not include individuals' ethnic background, thus this information had to be inferred from their names. Using this data, Sin (2002b, 2003) calculated the pair-wise indices of dissimilarity between the major ethnic groups and found that they were generally low, implying that ethnic segregation was low during the years studied.

Sim et al. (2003) adopted a similar name-to-ethnicity matching approach, but using a database of names from the Tanjong Pagar-West Coast Town Council, which contains individualised data on all household heads in the constituency who have registered for a HDB flat. Their study found that while the between-group indices of dissimilarity were considered low, the spatial distribution of the ethnic groups indicated certain subzones that had disproportionately more members of one ethnic group.

More recently, Wong examined the impact of the EIP on housing prices (2012) and investigated how people's preference for living with own-ethnic-group neighbours was influenced by the EIP quotas (2013). She gathered individualised data by name-matching from 2005 and 2006 phonebook listings and obtained quota limit status from the HDB's website (we adopt the latter method in our own study and will elaborate on it in the next section). Further, Wong (2013) estimated the location preference by comparing the changes in address between the 2005 and 2006 listings. Wong (2013) found that all ethnic groups express some preference for living near co-ethnics, but only up to a certain level. Once a neighbourhood has too many co-ethnic neighbours, people would prefer residents from other ethnic background to move in rather than people with the same racial profile.

To our knowledge, there has not yet been a nationwide study that maps the residential ethnic proportions across Singapore. Sin (2002b, 2003)'s and Sim et al's. (2003) research, while geographically-based, were focused on particular zones, namely Bedok and Tanjong Pagar-West Coast respectively. Wong's (2012, 2013) method encompassed the whole of Singapore but did not incorporate the geographic aspect. In the current research we explore the spatial distribution of all HDB blocks afftected by the EIP quotas, with the aim to identify housing estates that have a higher concentration for an ethnic group. This is followed by an examination on the correlations between EIP quota and socio-economic profiles in the neighbourhoods.

While past studies (Sin 2002b, 2003; Sim et al. 2003) have used statistical metrics such as the index of dissimilarity to estimate segregation, we refrain from doing so as such indices are influenced by how the boundaries are drawn. Instead, we studied the spatial locations of all blocks affected by EIP quotas and its association to the immediate environment using secondary data of the neighbourhoods where the blocks are located. Collectively, the locations of these affected blocks and the characteristics of their respectively neighbourhoods (e.g., age, density, income, and resale prices of HDB flats) will inform policymakers about the underlying racial and socio-economic dynamics within and across housing estates. In summary, there are two objectives for the current study. First, to understand the existing state of ethnic residential concentration in Singapore using EIP as a proxy indicator, and two, to find out if racial clustering is systematically linked to any form of socio-economic attributes.

4 Method

First, we obtained the list of HDB blocks from the API that is available on HDB Map Services.[1] For each HDB block identified by its postal code, we then checked its current EIP status for each racial group, using HDB's service for checking the buyer's eligibility under the EIP portal.[2] Out of the 8442 HDB blocks that we screened from June to July 2016, 1497 have hit the maximum quota for Chinese, while 399 and 574 units have hit the Malay and Indian/Others quota, respectively. It should be emphasised that the quotas are interlocked (i.e., having more of one race means less of another) but statistically it is possible that a HDB block to have reached two quotas (e.g., reach maximum ethnic quotas for Chinese and Indian, or Chinese and Malay, or Malay and Indian). Indeed, 88 blocks have hit the ethnic limit for both Malay and Indian, 26 blocks for both Chinese and Indian, and one HDB block for both Chinese and Malay. For clarity and parsimony reasons, we will examine each of the three races independently.

To visualise the locations and distribution of these blocks, we used OneMap API[3] to retrieve the full information of the address by postal code. We also identified the subzone and planning area for every block, by joining point locations for each block with subzone data taken from the 2010 population census (see Fig. 1). To identify where are the potential ethnic clusters, we ran the location data through the Spatial Autocorrelation (Global Moran's I) and Cluster and Outlier Analysis (Anselin Local Moran's I) tools in ArcGIS.

In addition to both Global and Local Moran's I indices, we generated a random list of HDB locations using the probability of a block affected by EIP quota, but assuming no ethnic clustering. In other words, a random distribution of HDB blocks affected by each of the three races based on the probability of being affected by EIP ethnic limits, as follows:

- Probability (random block meets Chinese quota) = Total HDB blocks meet Chinese quota/Total HDB blocks = 1497/8442 = 0.18
- Probability (random block meets Malay quota) = Total HDB blocks meet Malay quota/Total HDB blocks = 399/8442 = 0.05
- Probability (random block meets Indian/Others quota) = Total HDB blocks meet Indian/Others quota/Total HDB blocks = 574/8442 = 0.07

The random dispersion of blocks affected by each of the three races can be found in Figs. 2, 3, and 4. They form the backdrop for visual inspections against the observed spatial locations of blocks affected by ethnic quotas.

To explore the correlations between the EIP quota, demographic profiles, and social-economic indicators, we overlaid the EIP status of all HDB blocks with the demographic data from various open source sites. The demographic data include: (i) distribution of elderly residents aged 65 and above (by subzone), (ii) population

[1] https://services2.hdb.gov.sg/web/fi10/emap.html.

[2] https://services2.hdb.gov.sg/webapp/BB29ETHN/BB29STREET.

[3] https://docs.onemap.sg/#search.

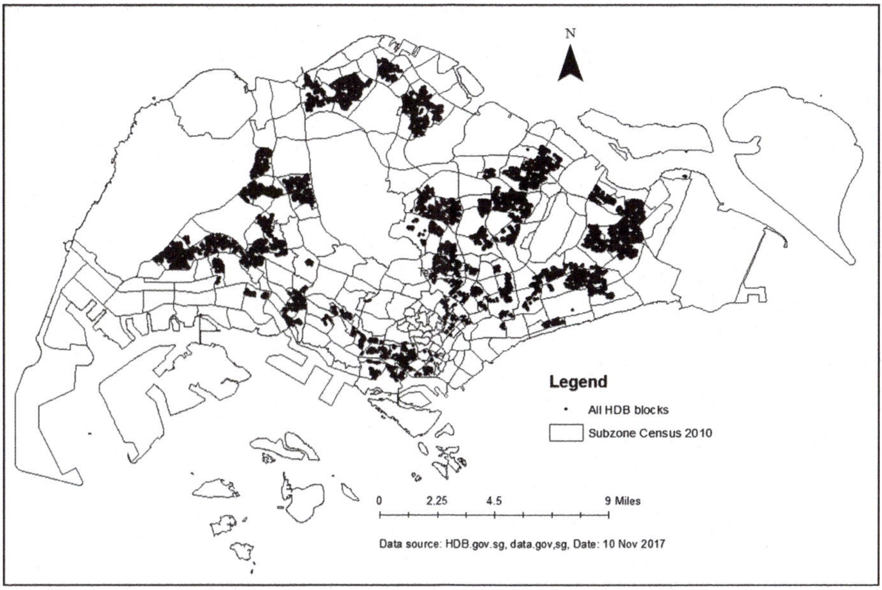

Fig. 1 Location of all HDB blocks

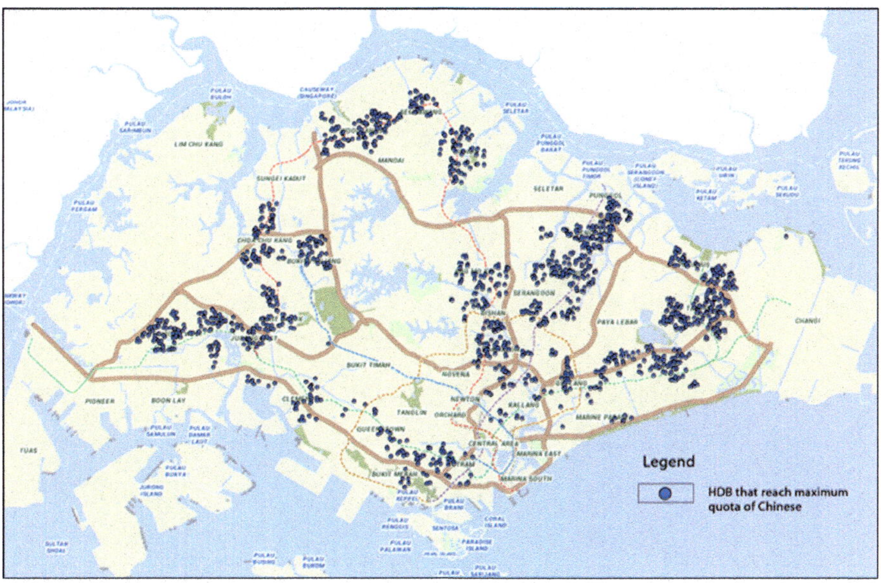

Fig. 2 Random distribution of HDB flats affected by Chinese quota

Fig. 3 Random distribution of HDB flats affected by Malay quota

Fig. 4 Random distribution of HDB flats affected by Indian quota

density (by subzone), (iii) percentage of households with at least $10,000 monthly income (by planning area which contains multiple subzones), and (iv) average resale price of all HDB blocks over a one-year period (November 2016 to Oct 2017; data available from HDB Map Services). The demographic data, including income, were retrieved from the 2015 General Household Survey. We automated the retrieval of HDB resale price data through a Ruby programming script, and obtained more than 37,000 transactions in total. The average resale unit price per square meter for each subzone provides a proxy indicator of socio-economic status.

5 Results

There are a total of 8442 HDB blocks, and out of these, 2355 blocks, or 28% of all public housing, reported reaching at least one ethnic quota limit; 1497 of them have hit the EIP Chinese quota, while 399 and 574 units have hit the Malay and Indian/Others quotas, respectively. The locations of the affected block for each of the three races can be found in Figs. 5, 6 and 7. The affected blocks for the respective racial groups are all located in distinct clusters, and the clustering is consistent with the Global Moran's Index values that range from 0.136 (Malays) to 0.454 (Chinese), with Z-scores well above 2.58 and p-values under 0.001.

The Chinese clusters are mostly situated in the central areas (e.g., Bishan), or in the city like Chinatown where the Chinese settled in traditionally, and at the periphery of

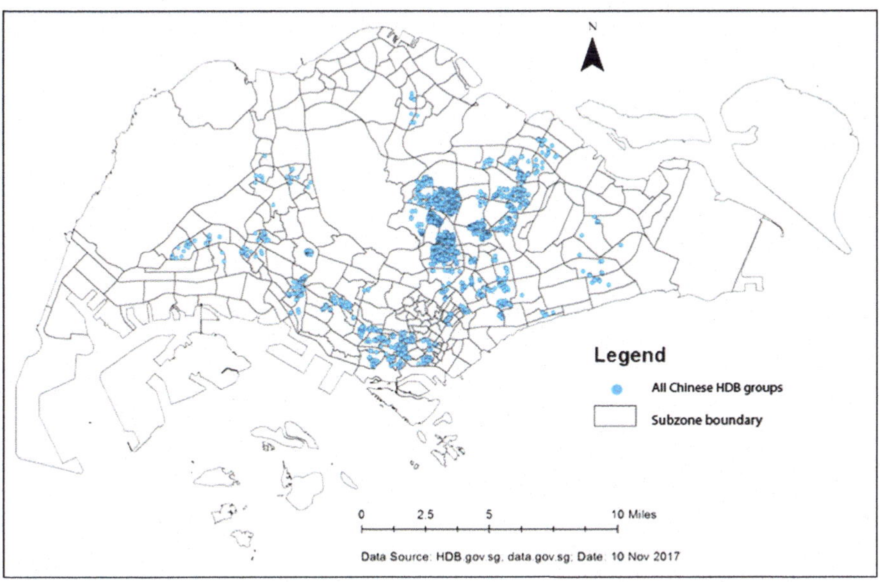

Fig. 5 HDB blocks that reached maximum Chinese quota

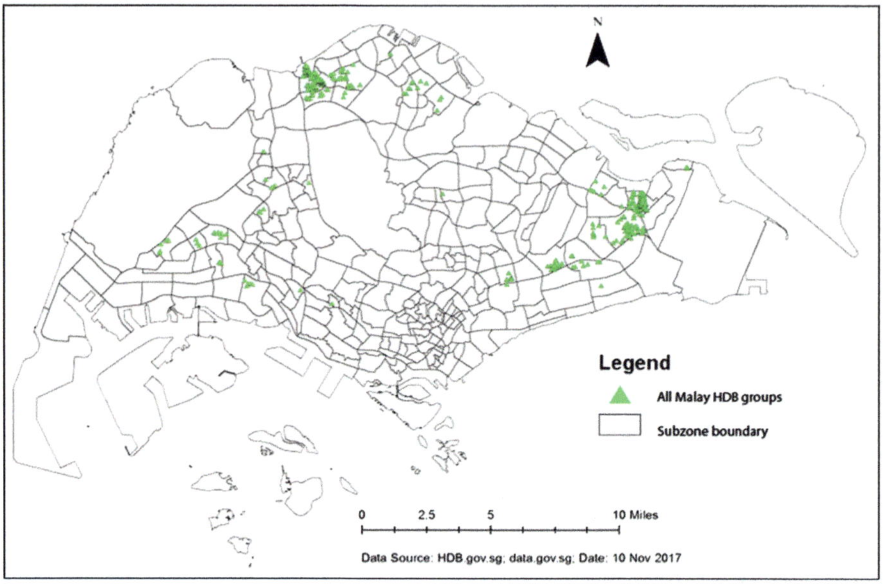

Fig. 6 HDB blocks that reached maximum Malay quota

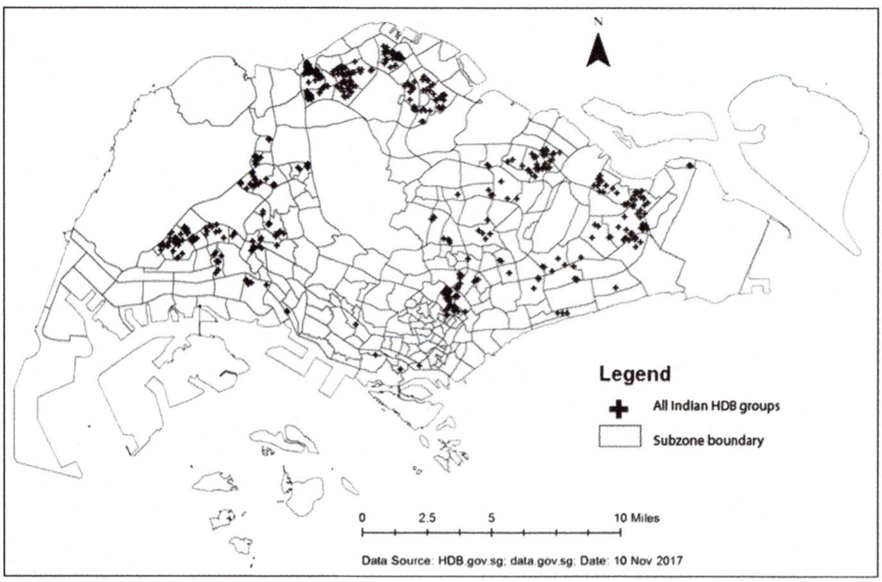

Fig. 7 HDB blocks that reached maximum Indian quota

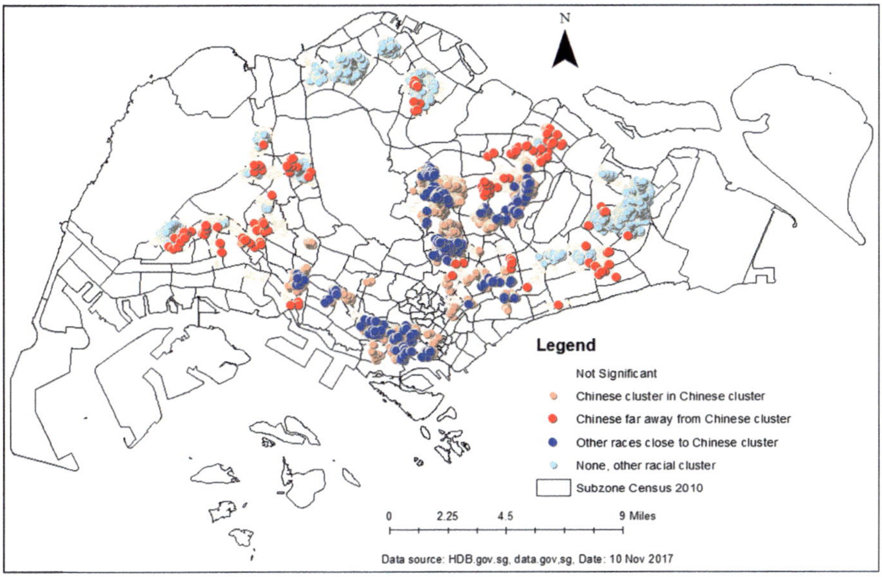

Fig. 8 Local Moran's I for HDB blocks that reached Chinese quota

the city (e.g., Tiong Bahru). The Malay clusters are found in fringe areas at northern and far eastern part of Singapore. The Indian/Other clusters are more balanced in comparison, as they can be found in fringe areas (northern, western, northeastern) as well as central regions and Serangoon, which is traditionally known as an Indian settlement. Comparisons between the actual locations of blocks affected by EIP, and a random dispersion of blocks affected by EIP demonstrate a sharp contrast, and hence nascent signs of enclaves (Chinese: Figs. 2 vs. 5; Malay: Figs. 3 vs. 6; Indian/Others: Figs. 4 vs. 7).

Further analyses using Local Moran's I corroborated the conclusion. Local Moran's I is a tool to detect localised concentration of point patterns. Areas that are marked in brown reveal the locations of ethnic clusters (i.e., enclave), the ones marked in sky blue show the under-representation of the ethnic group, the ones in red indicate a member of the ethnic group located within a cluster of other races (i.e., outlier), and lastly, the ones marked in navy blue show a member of another race located in the current ethnic cluster (i.e., outlier). Figures 8, 9, and 10 demonstrate the Local Moran's I mapping for the affected Chinese, Malay, and Indian/Others blocks, respectively.

To account for the irregular distribution of HDB blocks within each subzone (some subzones are very large but with few blocks built within a small portion of the parameter), we tried to normalise the data for each subzone using the total number of blocks in that subzone. As a result, we were able to identify subzones that contain a significant number of blocks (at least 20) and are significantly affected by the EIP quota. Twenty-three subzones have at least 50% of blocks hit the maximum EIP quota

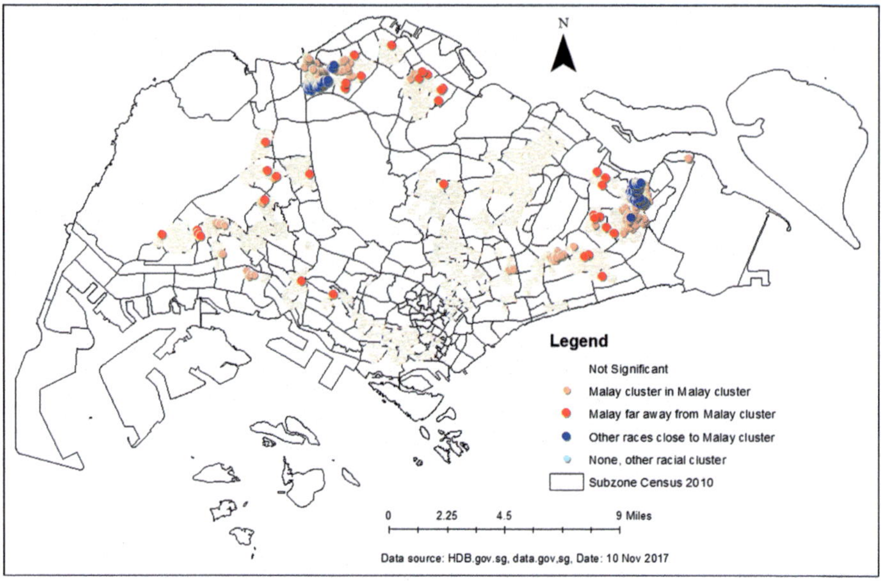

Fig. 9 Local Moran's I for HDB blocks that reached Malay quota

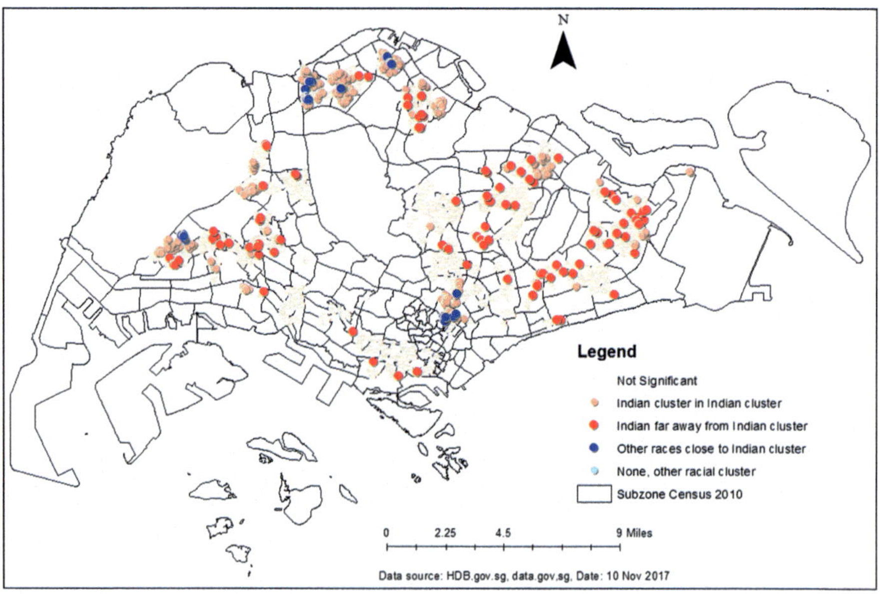

Fig. 10 Local Moran's I for HDB blocks that reached Indian quota

Table 2 Subzones where all HDB flats have reached at least one ethnic quota

Chinese	Malay	Indian/Others
Tiong Bahru	Woodlands West	Kampong Java
Holland Drive	North Coast	Admiralty
Upper Paya Lebar		North Coast
Marymount		
Serangoon Central		
Bishan East		
Chong Boon		
Cheng San		

Source Data retrieved between June and July 2016, from HDB online portal for ethnic eligibility (https://services2.hdb.gov.sg/webapp/BB29ETHN/BB29STREET)

for at least one of the racial groups; 13 of them have 100% of the contained blocks hitting the quota (see Table 2). In other words, the quota is completely filled up for the entire subzone for a specific racial group, for example, Bishan East/Marymount is "full" for the Chinese, while Woodlands West and Kampung Java are "full" for Malays and Indians/Others respectively. The subzone in North Coast is "full" for both Malays and Indians/Others.

We overlay the spatial locations of blocks affected by EIP quotas with subzone and planning area data on percentage of elderly residents aged 65 and above, population density, household income (demarcated at planning area), and the average resale price per square metre for HDB apartments (see Figs. 11, 12, 13 and 14). We found a visual correlation between the subzones with higher proportion of elderly residents and locations of EIP-affected blocks (Fig. 11). This is likely due to the maturity of the HDB flats, as the estate and its residents age together, and the fact that these estates have been around longer and thus have more time for people of the same race to congregate and eventually reach the EIP limit, if they happen to find the estate to be particularly desirable (Fig. 12).

The overlay between EIP status and household income is less obvious (Fig. 13), as clusters exist in both the less expensive fringe areas, as well as the more expensive central areas. However, there is seemingly a strong association between Chinese clusters and mean resale price per square metre (Fig. 14); ethnic minorities are more likely to cluster on the fringe areas with lower HDB resale prices, but the Chinese have a distinct presence in central areas which tend to command a higher resale price for HDB flats. The Malays are more likely to settle further away from the city. Apart from the historical presence of the Chinese in Chinatown (central), and the Malays in Eunos/Changi (central-east to far-east), it seems that the Chinese tend to concentrate in relatively more expensive central areas.

The statistical correlation at the subzone level between each of the three demographic indicators found that locations with higher concentration of ethnic Chinese clusters have significantly more elderly residents, lower density, and higher HDB

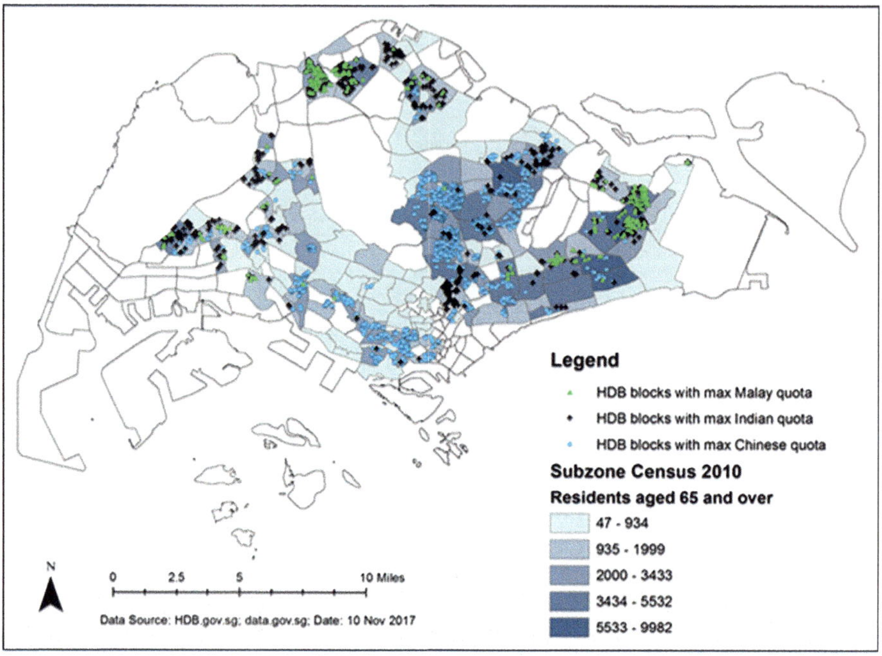

Fig. 11 Elderly residents and HDB blocks that reached EIP quota

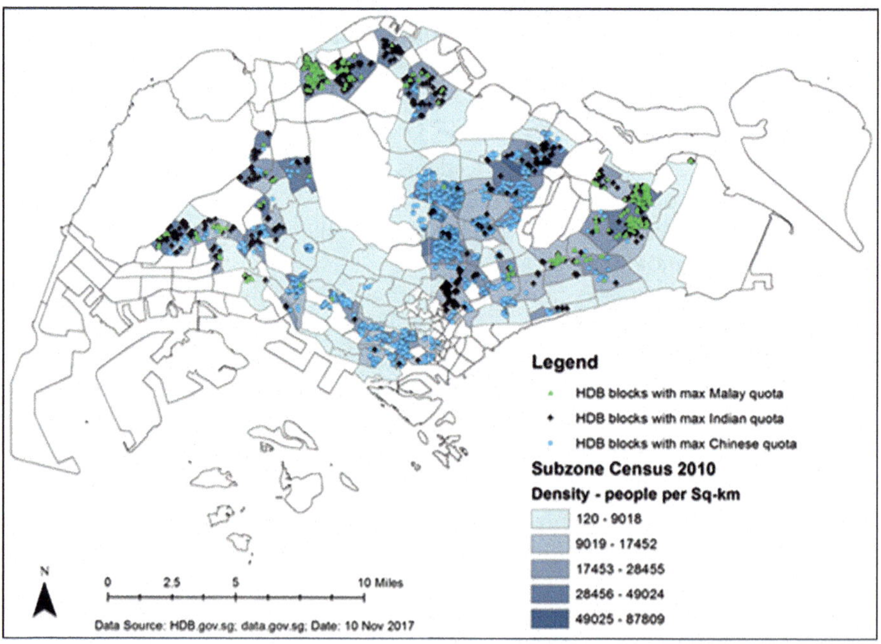

Fig. 12 Residential density and HDB blocks that reached EIP quota

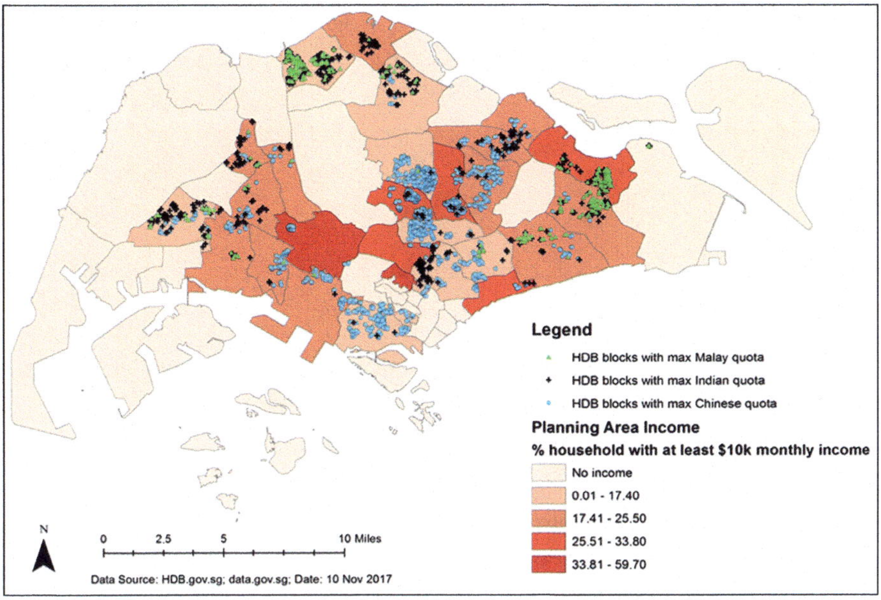

Fig. 13 Household income of planning areas and HDB blocks that reached EIP quota

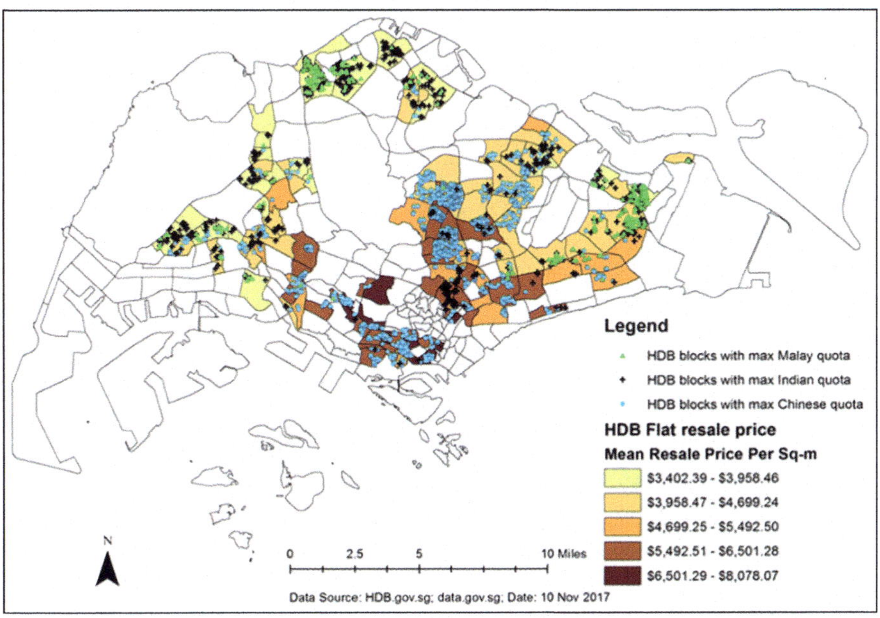

Fig. 14 Average resale price per square-metre and HDB blocks that reached EIP quota

Table 3 Correlation between flats affected by EIP limits and socio-demographic indicators

	Age 65 and above	Residential density	HDB resale housing price ($\$/m^2$)	Household income
% flats that reach Chinese quota	0.463***	−0.216*	0.597***	0.330*
% flats that reach Malay quota	−0.104	−0.033	−0.216*	−0.055
% flats that reach Indian/Others quota	−0.046	−0.214*	−0.020	0.026

*$p < 0.05$, **$p < 0.01$, ***$p < 0.001$

Source Department of Statistics Singapore (2017) and HDB online portal for ethnic eligibility (June to July, 2016)

resale prices; locations with higher concentration of Malay have lower resale price; whilst locations with higher concentration of Indian/Others clusters have lower density (see Table 3). The correlation for national income at the level of planning area shows a strong, positive correlation between higher concentration of Chinese residents and household income.

6 Discussion

This study examines the utility of EIP as a policy vehicle to mitigate the formation of racial enclaves in Singapore public housing estates. More specifically, the research aims to find out if there is a spatial dimension to HDB blocks that exceed a given racial quota; in other words, do apartment blocks that exceed a racial quota (e.g., Chinese) cluster around selected neighbourhoods? What are the socio-demographic characteristics of the clustered neighbours? The spatial analyses found significant clustering effects for each of the three races. Ethnic Chinese are concentrated in the central parts of Singapore, in subzones such as Cheng San, Chong Boon, Marymount, Bishan East, and the mature suburbs close to the city-central such as Tiong Bahru, Bukit Ho Swee, and Holland Drive. In contrast, there are disproportionately more Malays located around the northern and eastern parts of Singapore, and in subzones such as North Coast, Woodlands West, Tampines East, and Pasir Ris Drive. Ethnic Indian clusters are observed in subzones such as Admiralty, Midview, Seng Kang town central, Yunnan, Wenya, Boon Lay, Kampong Java, and Bendemeer. In total, 13 subzones have 100% of their HDB blocks reached the maximum ethnic quota.

Statistical analyses for global clustering (Global Moran's I) in each of the three ethnic categories are found to be significant, and it shows that there is a geographic concentration of ethnic groups across the island. This evidence is further corroborated by the Local Moran's I analyses for each racial group, as it reveals the specific locations with a high concentration of each racial group. The finding is consistent with the anecdotal observations. The current research takes one step further to examine if the clusterings are associated with any socio-demographic factors.

When we overlay locations of HDB blocks affected by racial quota against the demographic characteristics of each subzone (i.e., percent of elderly residents aged 65 and above, population density, and average HDB resale price per square metre) and planning area (household income), a portrait of racial divide has surfaced, though the cause and effect cannot be concluded. Residential subzones with a Chinese concentration tend to command higher HDB resale prices, receive higher household income compared to other planning areas, demonstrate lower housing density, and have a larger proportion of residents aged 65 and above. Neighbourhoods with a higher concentration of minorities are linked to lower HDB resale values. Resale home prices, household income, and lower density are proxies of socio-economic status, and the findings point to a potential class divide along racial lines in public housing. It is a sign that the Singapore racial polity has become more complex. It should also be stressed that there are many explanations to the correlation between ethnic clusters and social class—for instance, proximity to places (e.g., temples, mosques) that are familar to the different ethnic groups is an incentive that shapes housing preferences. In-spite of the current findings, EIP plays an important role in ensuring that no public housing estates in Singapore will become a complete racial enclave.

6.1 Residential Dynamics: Income Effect Versus Racial Effect

The estates that had the ethnic over-representations in the 1980s—which led to the implementation of the EIP—remain largely similar in terms of racial profile of its residents. Hougang, Redhill, and Bukit Merah continue to have a disproportionate number of Chinese residents; Bedok with a significant concentration of Malay residents; and Yishun and Kampong Java with an overrepresentation of Indians/Others. Currently Jurong West and Seng Kang estates have a sizable proportion of Indian/Others households, whilst the ratio of Malay households in Pasir Ris has grown substantially.

One could argue that the EIP only arrested the then-growing trend of ethnic regrouping and kept it under manageable levels, but was never meant to recalibrate ethnic proportions across the island. The block and neighbourhood limits (see Table 1) for each ethnic group roughly correspond to the ethnic proportions in the estates highlighted by the HDB in 1989; as of 1988, Bukit Merah was 88.4% Chinese, Bedok was 24.5% Malay, and Kampong Java was 15.2% Indian/Others (Ooi 1993). The limits were possibly set at those levels to allow housing transfers through the secondary market to cool the ethnic regrouping over time.

However, the limits set do not explain how and why ethnic congregation still persists beyond national proportions, and why other areas with ethnic overrepresentation have sprouted up. On one hand, there could be a racial effect where people prefer to live in a neighbourhood that is similar to one's ethno-cultural background as co-ethnics can understand one's cultural needs better. This behaviour was precisely what the policy was intended to arrest, as there is evidence that having co-ethnic neighbours continues to be a significant factor in home buyers' choice of residential location (Wong 2013).

On the other hand, ethnic congregation could be persisting due to an income or class effects driven by resale prices. Malay residents earn a lower national median income, and consequently they have a smaller budget for resale home purchase and the choices for housing locations are therefore confined to less expensive neighbourhoods. Chinese residents however earn a higher median national income and thus have more options in terms of choice locations. The confluence of a market pricing mechanism and purchasing power disparity between the races reinforces residential ethnic clustering.

The collateral effect of the EIP quota restrictions on housing prices has been well-discussed (Sin 2002a; Wong 2013). Since in an ethnically-constrained block a seller can no longer sell to a buyer of the quota-limited ethnic group, their pool of potential buyers has essentially shrunk. In these situations, the seller would either have to take a longer time to find a suitable buyer, or be willing to accept a lower selling price, often below market rate. Because of Singapore's overwhelming Chinese majority, this problem tends to be more acute in Chinese-constrained estates (i.e. the quota for Chinese residents has been reached), as Chinese sellers can sell to anybody, but non-Chinese can only sell to non-Chinese buyers. In a minority-constrained estate, sellers are only restricted from selling to the minority group in question (Wong 2012). The impact to resale home pricing and how it shapes ethnic segregation warrants further research.

The inequality in housing resale prices would hit the economically disadvantaged the hardest, especially those who are also ethnic minorities. In 2003, the sales restrictions were said to pose a "serious financial risk to some families" (Singapore Parliamentary Debates 2003). According to data from the 2015 General Household Survey (see Table 4), there is a correlation between ethnic group and socio-economic status. The median monthly household income from work is lower for Malays than for the other groups, while Malays are also significantly overrepresented among HDB 1- and 2-room flat dwellers. Chinese on the other hand are overrepresented among private property dwellers. Since housing is also a form of investment for most Singaporeans, any sales restrictions in the resale market could have repercussions to financial well being of Singaporeans, especially among the lower income groups.

Also noteworthy is the significant rise in the average monthly household income of Indian households from 2000 to 2010. In 2000, the figure was SGD$4623, which was below the national average of SGD$4988. In 2010, their average income was SGD$7664—above the national average of SGD$7214. The uplift in household income among the ethnic Indian community in 2010 was attributed to the influx of the better-educated—and higher-earning—foreign-born Indians over the last decade (Basu 2015). It remains to be seen if the spatial concentration of Indians in estates with lower resale prices are predominantly local-born Singaporeans, and if this turns

Table 4 Dwelling type and media household income from work by ethnicity

	Total (%)	HDB 1- and 2-room (%)	HDB 3-room (%)	HDB 4-room (%)	HDB 5-room and Exec. flats (%)	Condos and Other apts. (%)	Landed (%)	Median household income from work (SGD$)
Chinese	76.5	59.6	75.9	75.7	77.1	**81.4**	**85.1**	$7000–$7999
Malays	10.9	**28.3**	12.6	13.2	10.5	1.5	1.4	$5000–$5999
Indians	9.2	10.9	9.2	8.6	9.8	9.7	7.1	$8000–$8999
Others	3.3	1.2	2.3	2.4	2.6	7.5	6.4	$12,000–$12,999

Note Figures in bold indicate overrepresentations
Source Department of Statistics 2016

out to be the case, the effectiveness of EIP will be further compounded by another layer of social complexity.

To encourage more even distribution of the races across the different neighbourhoods, the government could focus more on enhancing the pull factors such as physical amenities and infrastructure. Localised research into neighbourhoods of high ethnic congregation could determine what features attract certain races to want to live there. An in-depth study into the housing preferences of the different ethnic groups would provide insights to the challenges. The findings can then be applied to making residential areas more attractive to the ethnic groups that are underrepresented. For example, the Chinese parents are drawn to schools that offer mandarin enrichment programmes for their school going children. Having more of such schools in neighbourhoods where there is a higher percentage of ethnic minorities such as Woodlands, Tampines, and Jurong West will attract more Chinese residents to relocate to the estates, and consequently a more balanced dispersion of ethnic groups across Singapore.

7 Future Directions

With more than 80% of the population residing in public housing, and close to 90% of them being home owners, EIP plays a critical role to dispersing ethnic communities across the city-state. However, the concentration of racial groups in the remaining 20% private residential properties and the profiles and geographic spread of rental tenants cannot be overlooked. The spatial locations of private landed properties are seemingly clustered in the central and northeast districts (e.g., Bukit Timah, Upper Serangoon, Aljunied). Future research may scrutinise this cluster of private home dwellers to examine how the residents' profile influences social cohesion across geographic locations.

The data collected is an extension to existing social science research on spatial housing segregation among diverse ethnic categories. Although there are clear statistical and empirical evidence to demonstrate the presence of racial concentration, the study has yet to show that the ethnic clusters are systemically linked to societal ills, such as an erosion of social trust in the neighbourhoods or racial tensions. A comprehensive study on the social impact of segregation, and the effectiveness of EIP in mitigating the clustering effects would require the current data to be overlaid against a larger, evaluative survey measuring social climate like neighbourhood confidence. This issue deserve a more thorough empirical investigation.

Lastly, with more than half a million permanent residents (PRs), or about 15% of the total resident population, there is a political imperative to ensure that immigrants integrate into the larger community. In March 2010, and in response to the groundswell of discontent against the staggering influx of immigrants to Singapore, a new residential quota for PRs was introduced for HDB apartments. This policy was aimed to prevent immigrant enclaves, with emerging signs that new migrants from the same cultural or national background may have the tendency to congregate in selected neighbourhoods, thereby hampering interactions with native-born Singaporeans. To ensure there is adequate dispersion of PR households, the policy imposes a cap of no more than 8% of the units in each block of flat to be owned by non-Malaysian PR households (e.g., PRs from China, India, Philippines), and not more than 5% of households within a given precinct (an area that is spatially similar to the size of a subzone). Malaysian PR households are exempted from the quota as they are culturally similar to Singaporeans, and hence have assimilated well into the Singaporean mainstream society.

In summary, the usefulness of this measure in preventing PRs concentration remains to be seen. Nevertheless, from the perspective of policymakers, geographic quotas on race or nationality are the main tools to achieving a multi-ethnic, cohesive Singapore and this is in line with "the state's policy of nation-building based on multiracialism" (Lai 1995). At the very least, EIP and PR residential quotas have provided a strategic direction to managing intercultural contact. Without these prerequisite conditions, HDB blocks in numerous subzones across Singapore are likely to be occupied by people of an identical race or nationality. This form of segregation will not be condusive to foster cross ethnic interactions, and in the long run detrimental to national cohesion. The exsiting system is not perfect but at the very least it provided an insurance against complete segegration.

References

Basu R (2015, February 26). Make foreigners, PRs count in income data. The Straits Times. Retrieved from: https://www.straitstimes.com/opinion/make-foreigners-prs-count-in-income-data

Department of Statistics Singapore (2015) Singapore Census of Population 2010, Retrieved from: www.singstat.gov.sg

Department of Statistics Singapore (2017) Population Trends, Retrieved from: www.singstat.gov.sg

Data.gov.sg (2017) Singapore Residents by planning area and type of dwelling, June 2016, Retrieved https://data.gov.sg/dataset/singapore-residents-by-planning-area-and-type-of-dwelling-June-2016

Lai AE (1995) Meanings of multiculturalism: a case study of ethnicity and ethnic relations in Singapore. Oxford University Press, Kuala Lumpur

Mohan M (2018, October 2) More appeals made for waiver of HDB's ethnic integration policy. Channel News Asia. Retrieved from https://www.channelnewsasia.com/news/singapore/hdb-ethnic-integration-policy-more-appeals-for-waiver-10780666

Mutalib H (2004) Singapore's quest for a national identity: the triumphs and trials of government policies. In: Choon BK et al (eds) Imagining Singapore. Eastern Universities Press, Singapore, pp 54–80

Ooi GL (1993) The housing and development board's ethnic integration policy. In: Ooi GL, Siddique S, Soh KC (eds) The Management of Ethnic Relations in Public Housing Estates. Times Academic Press, Singapore, pp 4–24

Phang SY, Kim K (2013) Singapore's housing policies: 1960–2013. In: Frontiers in development policy: innovative development case studies. Research Collection School of Economics, Singapore Management University, pp 123–153

Siddique S (1993) Ethnic relations and grassroots organizations. In: Ooi GL, Siddique S, Soh KC (eds) The management of ethnic relations in public housing estates. Times Academic Press, Singapore, pp 42–60

Sim LL, Yu SM, Han SS (2003) Public housing and ethnic integration in Singapore. Habitat International 27(2):293–307

Sin CH (2002a) The quest for a balanced ethnic mix: Singapore's ethnic quota policy examined. Urban Studies 39(8):1347–1374

Sin CH (2002b) Segregation and marginalization within public housing: the disadvantaged in Bedok new town Singapore. Housing Studies 17(2):267–288

Sin CH (2003) The politics of ethnic integration in Singapore: Malay 'regrouping' as an ideological construct. Int J Urban Reg Res 27(3):527–544

Singapore Parliamentary Debates (2003, November 10). Official Report vol 76. Singapore

Singapore Parliamentary Debates (2010, March 5) Official Report vol. 86. Singapore

Van Grunsven L (2000) Singapore: the changing residential landscape in a winner City. In: Marcuse P (ed) Globalizing cities—a new spatial order. Blackwell, Oxford, pp 95–126

Wong M (2012) Estimating the impact of the ethnic housing quotas in Singapore. University of Pennsylvania, Real Estate Department Working Paper

Wong M (2013) Estimating ethnic preferences using ethnic housing quotas in Singapore. Rev Econ Stud 80:1178–1214

Yuen B (2007) Squatters no more: Singapore social housing. Global Urban Development Magazine. Retrieved from: http://www.globalurban.org/GUDMag07Vol3Iss1/Yuen.htm

Building Social Resilience Through Parks and Common Recreational Spaces

Angelia Sia, Ee Heok Kua and Roger Ho

Abstract There is an increasing body of research that validates the connection between the quality of the physical environment and human health. In particular, the provision of public parks that are easily accessible to city dwellers is considered highly valuable. Green open spaces, in addition to their environmental benefits, have the potential to increase people's physical activity levels. With apt designs, parks offer a multi-sensorial environment that stimulates and improves mental health. These multi-tasking spaces also bring about social benefits; they provide opportunities for users of urban parks to interact with other users, and for users to become attached to the area (place attachment). Social interaction and place attachment are thought to contribute toward social cohesion, a collective identity and community support. These are all characteristics of resilient cities. Singapore's National Parks Board has undertaken a series of research studies in collaboration with medical professionals, that seek to understand these important aspects of social resilience in cities.

Keywords Social resilience · Urban greenery · People-nature interaction · Therapeutic gardens · Horticultural therapy · Therapeutic horticulture · Well-being · Ageing

1 Introduction

The City Resilience Framework describes city resilience as the capacity to function, in that urban dwellers can survive and thrive no matter what stresses or shocks they encounter (The Rockfeller Foundation & ARUP 2015). The original emphasis in the concept of resilience had leaned toward the general persistence of ecological

---

A. Sia (✉)
Centre for Urban Greenery and Ecology Research, National Parks Board, Singapore, Singapore
e-mail: angelia_sia@nparks.gov.sg

E. H. Kua · R. Ho
Department of Psychological Medicine, National University of Singapore, Singapore, Singapore
e-mail: pcmkeh@nus.edu.sg

C.-H. Leong and L.-C. Malone-Lee (eds.), *Building Resilient Neighbourhoods in Singapore*, Advances in 21st Century Human Settlements,
https://doi.org/10.1007/978-981-13-7048-9_4

system functions (Holling 1973). Over time, this definition has evolved to include two other characteristics, the adaptation of humans to nature (Holling 2001; Gunderson and Holling 2001; Berkes and Ross 2013), and their transformation in the face of global environmental changes. Social resilience was later framed in the context of individuals and communities, in regard to social and health challenges (Hegney et al. 2007; Buikstra et al. 2010). Social resilience involves recognising 'protective factors' (Buikstra et al. 2010) and building on existing strengths a community needs to meet adversity. This extension of resilience to include the social aspects is only logical; after all, environmental problems cannot be addressed in isolation from the social context (O'Brien et al. 2009).

The environmental benefits of urban nature—improved air and water quality, energy savings, as well as reduced urban heat island effects—has been extensively researched. More recently, social sciences provide a new dimension in the understanding of nature's benefits—its importance to human functioning and health. Given that 100% of Singapore's population lives in urban areas, it is important to create healthy urban environments for work, worship, learning, play, and an overall improved quality of life. National Parks Board (NParks), the government agency set up to provide and enhance the greenery of Singapore, has started a new and exciting journey to create a network of gardens as health-promoting spaces for everyone. This initiative is supported by a series of research studies carried out in collaboration with medical professionals, and complemented by other studies that seek to understand people's perceptions and behaviour in parks.

This chapter discusses how the tropical city-state Singapore builds social resilience through its parks and recreational spaces. Section two presents international research-based evidences that explain why humans may experience improved well-being when they interact with nature. This is followed by sections three and four, which introduce the parks provision and greenery-related initiatives in Singapore. The role of Singapore's greenery has also been expanded to promote human health, and this is discussed in a research case study in Sect. 5. The conclusion recapitulates the importance of urban parks and greenery in promoting individual's well-being and social resilience in cities.

2 The Role of Urban Greenery on Human Health

Urban planners recognise the importance of building resilient cities, and it is also common for cities to put in place greening plans that seek to integrate nature in the physical environment. Having a planned network of public green open spaces provides opportunities for people to interact with nature. This people-nature interaction is an important 'protective factor' that helps city dwellers recover from adversity or stress.

The restorative effects of people-nature interaction have been widely studied. A notable research was conducted by Roger Ulrich in the 1980s. In his study, he established that patients who were exposed to a view of nature recovered faster from surgery and needed a lesser amount of pain medication compared to patients who were exposed to a view of a brick wall (Ulrich 1984). Ulrich and his co-researchers further demonstrated that subjects who first watched a stressful movie and later exposed to colour/sound videotapes of natural settings experienced faster and more complete physiological recovery from the stressful movie, compared to those who viewed videotapes of urban settings after watching the stressful movie (Ulrich et al. 1991). Subsequent studies demonstrated and reinforced the benefits of people-nature interaction, and deepened the understanding of the link between nature, resilience and health. One noteworthy research suggested that views of natural elements such as trees, vegetation, and plants buffered the negative impacts of job stress (Leather et al. 1998).

The overarching theory that supports the association between people and nature is conceivably "biophilia". The biophilia hypothesis suggests that humans have an innate and biological need for connection to nature and its representations (Wilson 1984). According to this hypothesis, the tendency to connect with nature had enhanced the fitness of our ancestors, resulting in their brains becoming more attuned to extracting, processing, and evaluating information from the natural environment (Wilson 1984, Kellert and Wilson 2013). They acquired a compound system of rules learning in their interactions with nature, and these stimulations from nature had brought physical, psychological, and cognitive benefits.

In the modern day, humans are gradually distanced from the natural environment. Some researchers believed that the biophilic learning rules still prevail from generation to generation (Wilson 1993), but others cautioned that the biophilic instinct may have weakened in the lack of exposure to complex stimulations from nature-based interactions.

Considerably, the careful integration of nature into the urban environment is a worthwhile strategy that will contribute towards building and strengthening the resilient functioning of human populations. To maximise its benefits, the integration of nature in urban settings should be carefully planned and incorporated in landscape policies to boost and sustain human-nature interactions.

3 Provision of Parks and Landscapes in Singapore

Singapore's greenery and landscape policy began in the early 1960s, during its self-governance and rapid industrialisation period. The then-Prime Minister Lee Kuan Yew mooted the idea of greening the city to achieve the best possible living environment, and to provide a balance to rapid industrialisation and urbanisation (National Parks Board 2015). Singapore's greening programme started with the planting of large trees with big canopies along roadsides to provide shade for people.

In 1971, Singapore's Concept Plan specified a recreational open space standard of 0.13 ha per 1000 population at the district level (Gwee 2012). Large public parks, such as the East Coast Park and Mount Faber Park, were built at the time. They are popular destinations for Singapore residents, especially over the weekends.

To ensure that there are sufficient green spaces for the population to enjoy and benefit from, the recreational open space provision standard has been immensely raised, to the current 0.8 ha per 1000 people.

Today, NParks manages a network of over 350 parks and 300 km of park connectors. The park system comprises a range of green spaces, from large regional parks to small neighbourhood parks, all designed to meet the different recreational needs of the population. In fact, the urban scape in Singapore, too, has transformed to one where greenery is a major component.

4 Social Values of Parks and People-Plant Interaction

Besides their recreational role, parks function as spaces for social interactions, and facilitate social cohesion (Kuo et al. 1998). Park users in Singapore often consider parks as places to exercise, socialise and connect with nature (Yuen 1996). Large events, such as the Singapore Garden Festival, the Rockestra® concert series and the Ballet under the Stars, are being organised throughout the year, drawing masses to parks, enabling social interaction on a larger scale.

In Singapore, where the majority of its people dwell in high rise apartments, interactions among residents was found to be incidental and minimal (Cho and Ho 2014). Parks in the neighbourhood thus provide opportunities for people to meet and interact. Facilities in parks, such as playgrounds, were found to be particularly helpful as venues that promote socialising. This is consistent with the finding from the United Kingdom, which suggested that children play an important role in bringing people together and creating opportunities for parents at the playground to interact with one another (Kaźmierczak 2013).

NParks also promotes social interaction through gardening activities outside of parklands. The national Community In Bloom (CIB) programme has facilitated over 1300 community gardens in residential estates, schools and organisations.

5 Health Effects of People-Plant Interaction—Case Study in Singapore

More recently, NParks started a new initiative to create a network of health-promoting therapeutic gardens. This was in response to the recognition of the role of greenery in promoting human functioning and health. The purpose-built therapeutic gardens are designed to enable people with different levels of physical capability to engage in gardening as well as enjoy the restorative effects of nature. The gardens comprise both an activities zone, which incorporates universal design elements to cater to the needs of those who may be physically frail or differently able, as well as a restorative zone, which provides a plant dominated rehabilitative environment to provide relief from stress and promote mental well-being.

The idea of providing an enabling garden like environment is grounded on the scientific literature that engagement in horticultural tasks associates positively with better health and wellbeing. In a longitudinal study in New South Wales, Australia, researchers found that daily gardening predicted a lower risk of dementia (McCallum et al. 2007). To investigate if the health effects of gardening based programs were similar in tropical Singapore, NParks embarked on a number of interventional research studies of the elderly citizens, in collaboration with clinical researchers in the field of psychological medicine from the National University of Singapore (NUS).

Earlier in section two, we noted from Ulrich's findings how nature plays an important role in facilitating recovery from stress. The concept that underscores this benefit is based on the restorative characteristic of nature, expounded in the Attention Restoration Theory (ART), where "directed attention" is contrasted with "indirect attention". "Directed attention", an intentional response, is described as the attention needed by people to differentiate important and unimportant stimuli. In comparison, "indirect attention", sometimes also known as "fascination", is an automatic response that does not require much effort. Exertion of "directed attention" over a prolonged period is exhausting. It is perceived as a global inhibitory mechanism; the fatigue experienced from one task impacted other tasks that also require directed attention, and consequently impaired in an individual's mental competence (Kaplan 1995). The ART further elaborated that restoration of fatigued directed attention needs to be addressed, and that spending time in restorative environments provides a means of restoring this vital capacity (Cimprich 1993; Hartig et al. 1991; Kaplan 1983; Tennessen and Cimprich 1995).

The ART is often regarded as the foundation illuminating the restorative nature of people-plant interactions. Such interactions may be passive or active. Passive people-plant interactions encompass "viewing nature" and "being in nature". While "viewing nature" is self-explanatory, "being in nature" refers to both intentional and circumstantial encounters, such as wilderness experiences, visiting gardens, or incidentally walking past landscapes. Active people-plant interactions, on the other hand, are the direct and active engagements with plants through activities such as gardening (Kiyota 2009). A number of research studies have established that active people-plant interactions are vastly more effective in giving people a sense of purpose and meaning.

The benefits of active people-plant interaction were popularised after Dr. Benjamin Rush, "Father of American Psychiatry", published the book 'Medical Inquiries and Observations upon Diseases of the Mind', where he wrote that "digging in the soil seemed to have a curative effect on the mentally ill" (Rush 1812). It has also been documented that agricultural and gardening activities were consecutively included in both public and private psychiatric hospitals in the United States and used in the care of World War I veterans in their occupational and recreational rehabilitative treatment programmes (Detweiler et al. 2012). The use of plants in rehabilitative programmes later became more structured, and was termed horticultural therapy, the engagement of a client in horticultural activities to achieve specific and documented treatment goals. The practice of horticultural therapy started in the 1950s and was incorporated into the curriculum of an undergraduate programme in Kansas State University. One of the earliest advocators of horticultural therapy is the American Horticultural Therapy Association (AHTA). Founded in 1973 with the original name 'National Council for Therapy and Rehabilitation through Horticulture', AHTA assumed its current name in 1987. Widely considered to be the international authority of horticultural therapy, AHTA envisions to uphold horticultural therapy practice standards, and actively promotes research, training and professional certifications (American Horticulture Therapy Association 2018).

Extensive research-based evidence has been amassed, demonstrating the multiple benefits of horticultural therapy, from reducing chronic pain, improving attention, lessening stress to reducing falls. The studies were mostly conducted in temperate countries and in western societies. Hence, its effectiveness in the tropical environment had yet to be verified. To bridge the theoretical gap, NParks and NUS, in 2016, initiated the study to investigate the effects of therapeutic horticulture on the mental well-being of Asian elderly citizens. In the study, 69 elderly citizens between the age of 60 to 85 years old were recruited and randomly assigned to either the treatment or waitlist control group. The intervention was a gardening-based programme comprising 15 one-hour sessions. Conducted in a group setting, the treatment was termed therapeutic horticulture. The AHTA defined therapeutic horticulture as the participation in horticultural activities with the goal of enhancing participants' well-being, through active or passive involvement in plant and plant-related activities. The programme was facilitated by a registered horticultural therapist or other professionals who had received training in the use of horticulture as a therapeutic modality to support programme goals. (American Horticulture Therapy Association 2017).

The elderly citizens were selected as the study population due to Singapore's rapidly ageing population. Singapore has turned from a young society with only 2.5% of the population aged 65 and above in 1965 to the current 13% (Department of Statistics 2017). Its ageing projection was also pessimistic; the Singapore population is expected to age at a faster rate than that of other developed economies such as Australia, South Korea, the United Kingdom and the United States (International Longevity Centre, Singapore 2011).

The study hypothesised that there would be an improvement in the psychological well-being and general health of participants in the treatment group at the end of the therapeutic horticulture intervention, compared to the control group. The therapeutic horticulture sessions were designed to provide an enriching experience for the participants through interactions with nature. The range of activities aimed at promoting light physical activity and social interaction. There were also activities involving small nature art projects, and these were observed to promote a sense of self-worth, as well as to enhance the participants' attention span.

The study concluded with the main finding that the mean difference in scores on "positive relations with others" between the treatment and control groups was statistically significant. The participants in the treatment group observed a significant improvement in the scores on "positive relations with others", between the

Table 1 Mean score of "positive relations with others" at baseline and after six months

Allocated group	Baseline mean (SD)	6th month mean (SD)	Mean change (95% CI)	*P*-Value	Mean difference (Treatment vs. Control)	*P*-value
Treatment	15.07	17.21	2.14	.01*	2.84	.008**
	(3.60)	(2.99)	(.52 to 3.76)		(.75 to 4.92)	
Control	16.57	15.87	−.70	.311		
	(3.51)	(3.76)	(−2.09 to .69)			

Source Sia et al. 2018

pre- and post-intervention readings (Table 1). In comparison, a decrease in scores was observed for the participants in the control group.

Based on observations made by the research team, the therapeutic horticulture program was found to be effective in engaging participants constructively and promoting socialising. Participating in gardening in public parks also enabled other park users to see and enjoy the fruits of the participants' labour. In this respect, the elderly participants understood that their involvement in the treatment programme actually bettered the community, and hence had experienced a sense of purpose. Therapeutic horticulture is considered a meaningful activity that helps to build self-worth, confidence, and self-efficacy in a way that socialising on its own may not accomplish.

Besides psychometric assessments, there were also positive pre- and post-interventional changes observed in associated biomarkers (decrease in interleukin-6 and increase in CXCL12), suggesting that therapeutic horticulture offered biological benefits such as reducing inflammation and protecting neuronal functions in Singaporean elderly.

The research presented promising findings for NParks to continue with its plan to design parks for the purpose of nurturing human health. To build knowledge and skillsets on the practice of therapeutic horticulture, NParks has also organised specialised training programmes for healthcare professionals. These programmes were led by certified horticultural therapists and helped healthcare professionals develop treatment plans with specific goals that incorporated nature interaction.

Concurrently, specially designed therapeutic gardens in parks were planned and created. These gardens provide a plant-dominated environment that is purposefully designed to facilitate and enable interaction with the healing elements of nature. Therapeutic gardens have been completed in four parks, namely HortPark (Fig. 1), Bishan-Ang Mo Kio Park (Fig. 2), Tiong Bahru Park (Fig. 3) and Choa Chu Kang Park. These gardens have special provisions such as raised planter boxes, which enabled the elderly citizens to do gardening tasks without having to bend their backs (Fig. 4), and large shady trees to ensure that users are comfortable while performing gardening tasks (Fig. 5).

Fig. 1 Therapeutic Garden @HortPark. Credit Chelsea Sia

Fig. 2 Therapeutic Garden @Bishan-Ang Mo Kio Park. Credit Tham Xin Kai

6 Conclusion

The partnership between NParks and healthcare professionals marks the start of Singapore's journey in building social resilience through parks and recreational spaces. Through the local research study, the effectiveness of therapeutic horticulture on elderly citizens was validated. Furthermore, the benefits of therapeutic horticulture are being scaled up though the introduction of more specially designed therapeutic gardens and programmes across Singapore.

Fig. 3 Therapeutic Garden @Tiong Bahru Park. Credit Chuah Hock Seong

Fig. 4 Raised planter boxes enable elderly to engage in gardening without bending. Credit Chelsea Sia

The link between nature interaction and improvements in human health has been extensively demonstrated and accepted worldwide. Singapore, a City in a Garden, with its system of carefully planned and accessible parks and greenery, is well positioned to offer nature-induced well-being benefits to its people and visitors. Therapeutic horticulture, a recently introduced "active in nature" program, enables a deeper connection between people and nature in Singapore.

Fig. 5 Larger trees providing shade. Credit Chelsea Sia

Beyond therapeutic gardens in parklands, therapeutic horticulture may be easily designed and adapted as a complementary therapy to be used alongside conventional medical treatments to improve the health of specific populations like the elderly citizens, those with dementia, those recovering from stroke or those with neurodevelopment disorders. Plants and people share a rhythm of life. As people attend to their plants and gardens, they also nurture their bodies, minds and spirits. Hence, while one facet of greenery planning involves nurturing nature for environmental health, the other, which focuses on how nature can be adapted to nurture healthy population and sound individual health, is just as essential. With appropriate planning and design, parks and greenery, while functioning as common recreational spaces, multi-task to promote people's well-being and social resilience.

References

American Horticulture Therapy Association (2017, July) Definitions and positions paper. Retrieved April 26, 2018, from http://www.ahta.org/assets/docs/definitions and positions final 6.17.pdf

American Horticulture Therapy Association (2018) About the american horticultural therapy association. Retrieved November 5, 2018, from https://www.ahta.org

Berkes F, Ross H (2013) Community resilience: towards an integrated approach. Soc Nat Res 26:5–20

Buikstra E, Ross H, King CA, Baker PG, Hegney D, Mclachlan K, Rogers-Clark C (2010) The components of resilience-perceptions of an Australian rural community. J Community Psychology 38(8):975–991

Cho IS, Ho KC (2014) Impact of built environment on community bonding. Retrieved April 26, 2018, from http://hdb.gov.sg/cs/infoweb/doc/community-seminar-nus

Cimprich B (1993) Development of an intervention to restore attention in cancer patients. Cancer Nurs 16:83–92

Department of Statistics Singapore (n.d.) Population Trends 2017. Retrieved April 26, 2018, from http://www.singstat.gov.sg/docs/default-source/default-document-library/publications/publications_and_papers/population_and_population_structure/population2017.pdf

Detweiler MB, Sharma T, Detweiler JG, Murphy PF, Lane S, Carman J, Chudhary AS, Halling MH, Kim KY (2012) What is the evidence to support the use of therapeutic gardens for the elderly? Psychiatry Investigation 9(2):100–110

Gunderson L, Holling CS (2001) Panarchy: understanding transformations in human and natural systems. Island Press

Gwee J (2012) Case studies in public governance: building institutions in Singapore. Routledge

Hartig T, Mang M, Evans GW (1991) Restorative effects of natural environment experiences. Environment Behav 23:3–26

Hegney D, Buikstra E, Baker P, Rogers-Clark C, Pearce S, Ross H, King C, Watson-Luke A (2007) Individual resilience in rural people: a queensland study Australia. RuralRemote Health 7:620

Holling CS (1973) Resilience and stability of ecological systems. Annu Rev Ecol Syst 4(1):1–23

International Longevity Centre—Singapore (2011) A profile of older men and older women in Singapore 2011. Retrieved from https://tsaofoundation.org/doc/Profile_Of_Older_Men_-_Singapore.pdf

Kaplan S (1983) A model of person-environment compatibility. Environment Behav 15:311–332

Kaplan S (1995) The restorative benefits of nature: toward an integrative framework. Journal Environmental Psychology 15(3):169–182

Kaźmierczak A (2013) The contribution of local parks to neighbourhood social ties. Landscape Urban Plann 109(1):31–44

Kellert SR, Wilson EO (2013) The biophilia hypothesis. Island Press

Kiyota E (2009) People-nature interactions: the therapeutic role of nature in elderly residents' everyday experience in a long term care facility (Unpublished doctoral dissertation). The University of Wisconsin-Milwaukee

Kuo FE, Sullivan WC, Coley RL, Brunson L (1998) Fertile ground for community: inner-city neighborhood common spaces. Am J Community Psychol 26(6):823–851

Leather P, Pyrgas M, Beale D, Lawrence C (1998) Windows in the workplace. Environment Behav 30(6):739–762

McCallum J, Simons LA, Simons J, Friedlander Y (2007) Delaying dementia and nursing home placement. Ann N Y Acad Sci 1114:121–129

National Parks Board (2015) A special tribute to Mr Lee Kuan Yew—the man behind the greening of Singapore. Retrieved November 5, 2018, from https://www.nparks.gov.sg/about-us/special-tribute-to-mr-lee

Obrien K, Hayward B, Berkes F (2009) Rethinking social contracts: building resilience in a changing climate. Ecology Soc 14(2)

Rush B (1812) Medical inquiries and observations upon diseases of the mind. Kimber and Richardson

Sia A, Ng KS, Ng KW, Chan HY, Tan CH, Rawtaer I, Feng F, Mahendran R, Kua EH, Ho CM (2018). The effect of therapeutic horticulture on the psychological wellbeing of elderly in Singapore: a randomised controlled trial. J Therapeutic Horticulture 28(1)

Tennessen CM, Cimprich B (1995) Views to nature: effects on attention. J Environmental Psychology 15:77–85

The Rockfeller Foundation & ARUP (2015) City resilience framework. Retrieved April 24, 2018, from https://assets.rockefellerfoundation.org/app/uploads/20160105134829/100RC-City-Resilience-Framework.pdf

Ulrich RS (1984) View through a window may influence recovery from surgery. Science 224(4647):420–421

Ulrich RS, Simons RF, Losito BD, Fiorito E, Miles MA, Zelson M (1991) Stress recovery during exposure to natural and urban environments. J Environmental Psychology 11(3):201–230

Wilson EO (1984) Biophilia. Harvard University Press

Yuen B (1996) Use and experience of neighbourhood parks in singapore. J Leisure Res 28(4):293–311

Urban Mobility and Resilience: Transport Infrastructure Investment and the Demand for Travel

Siqi Song, Mi Diao and Chen-Chieh Feng

Abstract Urban mobility is a key ingredient to fostering social interaction and resilience. At the neighbourhood level, enhanced mobility can be achieved through investment in transport infrastructure, which in turn can promote greater community interaction and exchange, and in the process contribute to social resilience. This study examines the influence of transport infrastructure investment on household work and non-work travel demand in Singapore, with the main purpose of investigating whether improved transport accessibility brought by transport infrastructure investment prompts additional out-home trips; and if so, to what extent and in what way does transport infrastructure investment influence household travel demand. Our study indicates that the travel demand for both work and non-work purposes is mainly determined by households' socioeconomic and demographic characteristics. Rail transit accessibility and expressway proximity do not influence travel demand significantly. Land use mix and population density have a positive influence, though small in magnitude, on household travel demand, and consequently on mobility, which contributes to greater social resilience.

Keywords Urban mobility · Transport infrastructure investment · Travel demand · Neighbourhood resilience

---

S. Song (✉) · C.-C. Feng
Department of Geography, National University of Singapore, Singapore, Singapore
e-mail: siqi.song@u.nus.edu

C.-C. Feng
e-mail: geofcc@nus.edu.sg

M. Diao
Department of Real Estate, National University of Singapore, Singapore, Singapore
e-mail: rstdm@nus.edu.sg

C.-H. Leong and L.-C. Malone-Lee (eds.), *Building Resilient Neighbourhoods in Singapore*, Advances in 21st Century Human Settlements,
https://doi.org/10.1007/978-981-13-7048-9_5

1 Introduction

Singapore is a small island city-state with a land area of 719.9 square kilometres and a population of 5.61 million (year 2017 figure[1]), making it one of the most densely populated countries in the world. As with many other cities, Singapore has experienced a steady increase in travel demand spurred by its increasing population and rising affluence. Figure 1 shows the increase of travel demand for both private motor vehicles and public transport in Singapore over a number of years. Daily public transport ridership has more than doubled, from 3.97 million in 1996 to 8.17 million in 2016. In the meantime, the use of private motor vehicles has grown, although at a relatively slower pace. In 2005, mileage per vehicle was 196,139 km, and this value has increased to 207,600 in 2014, which exceeded US mileage per vehicle which was 18,702 km as of 2014.[2]

The Singapore government has been proactive in adopting various strategies in order to accommodate the fast growth of travel demand. One of such strategies is the investment in transport infrastructure. This supply-side management strategy encompasses two main components:

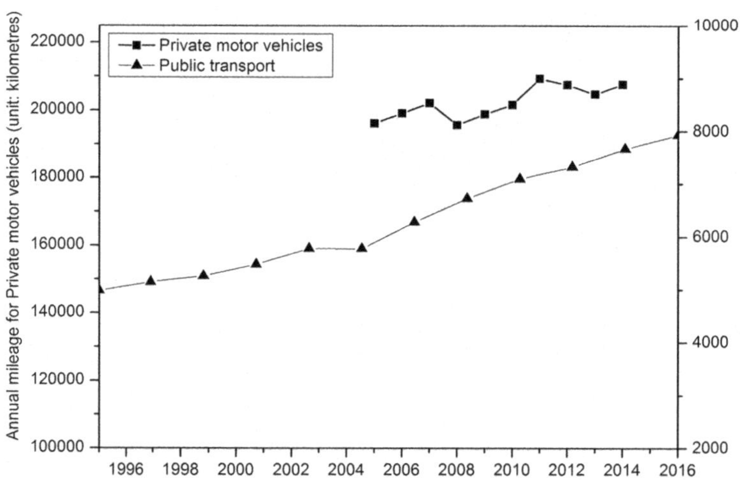

Fig. 1 Travel demand for private motor vehicles and public transport in Singapore from 1996 to 2016. *Source* Land Transport Authority, Singapore (Public transport includes Mass Rapid Transit (MRT), Light Rail Transit (LRT), bus and taxi. Private motor vehicles are cars, private hire buses, school buses, light goods vehicles, heavy good vehicle and motorcycles. Data retrieved from https://data.gov.sg/dataset/public-transport-utilisation-average-public-transport-ridership and https://data.gov.sg/dataset/annual-mileage-for-private-motor-vehicles)

[1]Data source: Department of Statistics, Singapore. Retrieved from https://www.singstat.gov.sg/statistics/latest-data#16.

[2]Data source: Department of Transportation, Federal Highway Administration, US. Retrieved from https://www.fhwa.dot.gov/policyinformation/statistics/2015/vm1.cfm. 1 miles = 1.609 km.

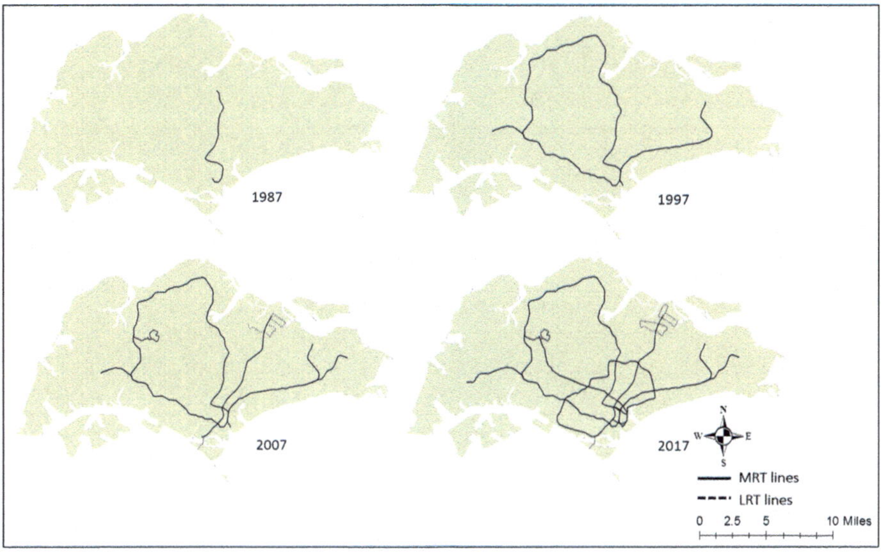

Fig. 2 Evolution of the transit network in Singapore

(i) **Rail transit construction**. Massive efforts have been made to develop and upgrade rail transit system (Fig. 2). The Mass Rapid Transit (MRT) network has been continuously expanded since it came into operation in 1987, with buses and the Light Rail Transit (LRT) serving as feeders to the MRT network. The first transit line, the North-South Line, was open to the public in November 1987. Three years later, the East-West Line which links the east and west part of Singapore was completed. Projects like North-East Line (2003), Circle Line (2009) and Downtown line (2013) have furthered the overall coverage of transit services across the island. Over the past decade, the length of the transit lines has been extended from 109.4 km in 2005 to 170.8 km in 2015.[3]

(ii) **Expressway expansion**. To meet the increasing vehicle travel demand, the government has dedicated to building more roads or expanding existing ones. New expressways and roads have been constructed steadily from 1986, but at a much slower pace as compared with rail transit development, largely due to land scarcity (Fig. 3). In 2007, Singapore's road network extended over 3297 km and now covers 12% of the total land area. It includes eight expressways with a total length of 153 km, compared with six expressways with a length of 129.1 km in 1990.

Rail transit construction and expressway expansion have induced significant changes in people's living environment, which are expected to influence their travel behaviour. This paper investigates such effect of transport supply. In particular, it

[3]Data source: Land Transport Authority, Singapore. Retrieved from https://www.lta.gov.sg/content/ltaweb/en/publications-and-research.html.

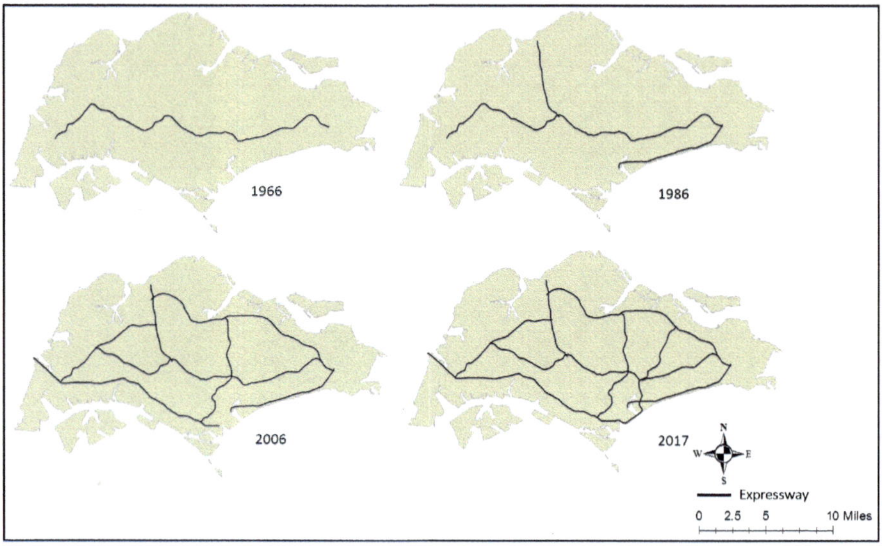

Fig. 3 Evolution of the expressway network in Singapore since 1966

examines the influence of improved transport accessibility induced by rail transit construction and expressway expansion on work and non-work household travel demand. We aim to answer the following questions: (1) Will transport infrastructure investment increase or suppress household travel demand? (2) If so, to what extent and in what way transport infrastructure investment influences household travel demand in Singapore. The results will help inform transport policy by providing further insights into how people's travel patterns are affected by the planned and designed physical environment.

Ideally, the analysis of the influence of transport supply on travel requires a longitudinal dataset (Mokhtarian and Cao 2008). However, longitudinal data are rarely available, especially in the transportation field. This paper, therefore, adopts a second-best approach, a pseudo-panel approach, which has been increasingly used by scholars (see e.g., Bernard et al. 2011; Dargay 2007; Meng et al. 2014; Tsai et al. 2013; Weis and Axhausen 2009). A pseudo panel is an artificial panel data, constructed by aggregating individuals of different cross-sectional surveys into cohorts with a consistent definition. The mean values of these cohorts are then treated as individual observations, which are followed over time. In this research, the pseudo panel is constructed based on three household travel surveys conducted in 2004, 2008 and 2012 in Singapore. Travel demand is represented by the total number of trips on a daily basis. Separate models are estimated for work trips and non-work trips given that the decision-making process is likely to differ for trips with different purposes.

The rest of the paper is structured as follows. The next section reviews relevant literature on transport supply and travel demand. Sections 3 and 4 introduce the data and the way to construct the pseudo panel, respectively. Section 5 presents the modelling approach, followed by the estimation results in Sect. 6. The last section summarises the key findings and provides relevant policy implications and future research directions.

2 Transport Supply and Travel Demand: A Literature Review

Travel is an important element in everyday life. Apart from enabling movement of people so that they can get to work and gain access to basic daily needs, such as grocery shopping, recreation and education, it also provides opportunities for people to partake in social activities, meet and interact with each other, and consequently contributes to neighbourhood or city cohesion and resilience. Travel demand, as a key aspect of travel decisions, captures people's travel needs, i.e., whether, how, for what purpose, and where people travel to. Greater mobility enhances the opportunity for people to meet and interact with neighbours, friends or even strangers. Thus, understanding how various factors affect travel demand is essential to maintaining the robustness of communities and their quality of life. This understanding is important for effective transport planning.

Existing studies on the relationship between the built environment and travel demand have assumed that trip rates tend to be higher in neighbourhoods with high density, mixed land use and easy access to public transport (Crane 1996; Ewing and Cervero 2001). This assumption is plausible as lower travel cost per trip is expected in areas where destinations are closer and more accessible.

Although theories imply that the trip rates are likely to correlate with the built environment around residence, empirical evidence from past studies on the relationship between travel demand and land use is far from conclusive. Some studies suggested that trip generation is mainly determined by household size, income or other demographic factors, and is inelastic to the built environment (see e.g., Ewing et al. 1996; Kitamura et al. 1997; Milakis et al. 2015), while others found a higher frequency of trip making in neighbourhoods with certain physical characteristics (see e.g., Boarnet et al. 2003; Frank et al. 2007; Kitamura 2009; Shay and Khattak 2012; Targa and Clifton 2005). Hence, whether and how urban planning influences people's trip-making decisions is still uncertain.

Past studies have used various kinds of measures to capture travel demand, which partially explains the contradictory research findings on the trip-making effect of the built environment (Ewing et al. 1996; La Paix Puello et al. 2013). In some studies, travel demand is represented by the total number of trips by all transport modes, while in others, it is the number of trips by a specific transport mode, such as walking, cycling or driving. Still, some studies use trip frequency differentiated by trip purposes or a combination of transport mode and trip purpose. For example, Boarnet et al. (2003) examined the influence of the built environment on individual non-work auto trips in Portland, US and suggested that income plays a more dominant role in non-work auto trip-making rather than land use attributes. Targa and Clifton (2005) analysed the effect of urban form, neighbourhood design and land use attributes on walking trip frequency with the 2001 National Household Travel Survey in the Baltimore metropolitan region. They found that people living in denser areas with higher street connectivity, easier access to transit or bus stops and mixed land use tend to walk more frequently. Milakis et al. (2015) investigated the role of both the local and regional urban form on vehicle travel measured by vehicle trip frequency and vehicle kilometres travelled in the San Francisco Bay Area. The results did not reveal any considerable influence of urban form on vehicle trip frequency. By comparing the travel behaviour of residents before and after the opening of a new rail transit line in Singapore, Zhu and Diao (2016) found a shift from private cars to rail transit in the corridor.

Moreover, most existing studies on travel demand are cross-sectional in nature, which does not allow researchers to draw a strong causal inference between the built environment and travel demand (Handy et al. 2005; Mokhtarian and Cao 2008). Longitudinal analysis has the potential to overcome this issue as it considers time precedence that is necessary for causality establishment. However, due to the lack of longitudinal data, such analyses aiming to explore the causal relation between built environment and travel demand have so far been limited to a few studies examining the effect of changes in the built environment on changes in activity frequency by mode (Troncoso Parady et al. 2014; Scheiner and Holz-Rau 2013). Our research overcomes the data limitation by employing a pseudo panel approach.

3 Data

3.1 Data Sources

The pseudo panel was created based on Household Interview Travel Survey (HITS) of Singapore. The Singapore Land Transport Authority (LTA) has carried out this survey once every four or five years since 1997. A total of around 10,000 households, nearly one percent of the total population, were interviewed each time. HITS consists of data at different levels of aggregation: household level, individual level and trip level, where household characteristics, personal characteristics and out-of-home trips on a random weekday were recorded, respectively. Data on the latter two were only collected from household members aged six and above. The detailed information on households, individuals and their travel makes HITS a usable database for travel behaviour analysis. In this study, we used HITS conducted in 2004, 2008 and 2012 to construct the pseudo panel.

Other built environment information such as land use, transport infrastructure supply was extracted from various publicly accessible online sources provided by LTA and the Urban Redevelopment Authority (URA). Specifically, the land use data were extracted from the Master Plan 2008 of Singapore.[4] Information on the island-wide transport network, transit lines and expressways, in particular, was obtained from the historical maps of transport systems in Singapore. Lastly, the population density information at the subzone level was collected from the census statistics provided by the Singapore Department of Statistics (DoS).

3.2 *Travel Demand Indicators*

Travel demand was represented by the total number of trips on a daily basis. Household was chosen as the unit of the analysis due to the potential trip trading and coordinated trip making among household members. The HITS provides detailed information on daily out-of-home trips (e.g., trip frequency, trip purposes and travel mode for each trip) made by each individual above the age of six in the households. This makes it possible to extract the data on the total trip frequency by trip purposes.

Travel behaviour varies considerably based on various trip purposes. Different activities are characterized by different degrees of choice and spatiotemporal constraints, depending on what needs the activity is intended to fulfill (Elldér 2014). For example, trips carried out to perform discretionary activities can reasonably be expected to have a more flexible relationship with spatial or time constraints than do more compulsory activities, such as work or grocery shopping. Individuals can generally choose more freely where and when to perform discretionary activities based on preferences, while labour-related trips have less freedom since work must be performed at particular workplaces and specific time. In this study, trip purposes were broadly classified into two categories: (i) work trips, which refer to all work-related activities, including going-to-work and trips serving for work purposes, and (ii) non-work trips that encompass all other activities, such as leisure, shopping, eating-out or exercising. A more detailed subdivision was originally planned for non-work trips. However, the fact that all the trips are counted on a daily basis leads to a relatively small count of non-work trips, making it meaningless to classify further. This altogether generated three travel demand measures:

(i) The total number of trips;
(ii) The total number of work trips; and
(iii) The total number of non-work trips.

3.3 *Transport Supply and Built Environment Variables*

Transport supply was measured by the distance to the nearest rail transit station and the nearest expressway exit, capturing the transport accessibility improved by

[4]For more information on the Master Plan 2008, see https://www.ura.gov.sg/maps2/?service=mp&year=2008.

Table 1 Key variables in the analysis

Variables (Unit)	2004	2008	2012
Travel-related variables			
Total number of trips (count)	6.175	6.158	6.378
Total number of work trips (count)	1.529	1.653	1.835
Total number of non-work trips (count)	4.645	4.505	4.543
Transport supply			
Distance to the nearest expressway exit (km)	3.537	3.444	3.363
Distance to the nearest LRT/MRT station (km)	1.436	1.309	1.260
Other built environment variables[a]			
Walking path density (km/km^2)	11.854	11.955	12.148
Land use entropy index	0.693	0.697	0.699
Population density (1000/km^2)	2.137	2.073	2.033
Socio-demographics			
Age of head of household	44.355	49.798	49.873
Household size (count)	3.786	3.261	3.410
Household income[b] (ordinal)	5.051	5.436	5.963
Car ownership (count)	0.336	0.385	0.341
Year dummy variables			
Yr2004	1	0	0
Yr2008	0	1	0

Notes
[a]Variables computed at the subzone level
[b]Income is classified into 12 categories. They are: no income; $1–1000; $1001–1499; $1500–1999; $2000–2499; $2500–2999; $3000–3999; $4000–4999; $5000–5999; $6000–6999; $7000–7999; and $8000 & above
Source Social-demographics are exacted from HITS, while other variables are computed by authors

rail transit construction and expressway expansion during the study period. These two variables were computed using the network analyst function in ArcMap 10.1 (ESRI 2010) for all three years under study, i.e., 2004, 2008 and 2012. Three other built environment variables were included as control variables, including density as measured by population density, diversity as represented by land use entropy index, and street connectivity as measured by walking path density. Table 1 provides a summary of all the variables included in this study.

3.4 Social Demographic Variables

Socio-demographics were included as control variables, as they are important determinants of people's travel behaviour. The selection of the socio-demographic variables was guided by previous travel behaviour research, but somewhat limited by the data availability and survey contents. The socio-demographics took into account individuals' or households' age, gender, income, vehicle ownership and household size, all of which are expected to influence the travel behaviour. Incorporating socio-demographics is also meaningful to mitigate the residential self-selection issue when attitudinal data are not available (Brownstone 2008; Brownstone and Golob 2009).

4 Construction of the Pseudo Panel

4.1 The Pseudo Panel Approach

The concept of pseudo panel was first introduced by Deaton (1985) in consumer economics to overcome the difficulties of acquiring real panel data. The pseudo panel is constructed based on existing repeated cross-sectional data and by grouping individuals into cohorts with variables that are or can be reasonably assumed to be time-invariant. Each cohort is then treated as a single observation, and the mean value of all individuals in each cohort is generated to represent each observation. Even though pseudo panel is not real panel, the cohorts can be treated as if they were the same persons and traced over time. The pseudo panel data allow researchers to perform longitudinal analysis at the defined-cohort level. Examples for the application of the pseudo panel approach in the transportation field are Dargay (2001), (2002), Weis and Axhausen (2009) and Tsai et al. (2013).

One limitation of the pseudo panel approach is that the estimation with pseudo panels is likely to suffer from a loss of individual information since group means are used. This problem can be alleviated if the heterogeneity between cohorts is maximized and the variation within cohorts minimized[5], which requires proper grouping criteria (Verbeek and Nijman 1992). The criteria used to group individuals into cohorts vary with the contexts of different studies. Birth year is the most commonly used variable as it is time-invariant and allows researchers to examine the lifecycle and generation effects (Tsai et al. 2014). Other complementary variables, such as gender or household location, are also used if necessary.

4.2 Grouping Criteria

In order to generate a pseudo panel that has sufficient level of disaggregation as well as large enough cohort size, we chose two grouping criteria:

[5] Heterogeneity/variation refers to the differences across the individual units being studied. A desired pseudo panel should have sufficient inter-cohort heterogeneity and within-cohort homogeneity to improve estimation efficiency.

Table 2 Household travel demand in 2004, 2008 and 2012 by Birth Year Groups

	Before 1945	1946–1955	1956–1965	1966–1975	After 1976
2004					
Total number of trips	4.441	6.746	7.276	6.047	6.363
Total number of work trips	0.959	1.645	1.478	1.709	1.856
Total number of non-work trips	3.482	5.101	5.798	4.338	4.507
2008					
Total number of trips	3.355	6.552	7.812	6.975	6.095
Total number of work trips	0.508	1.931	1.867	1.855	2.106
Total number of non-work trips	2.847	4.622	5.945	5.120	3.989
2012					
Total number of trips	2.992	6.726	8.259	7.085	6.829
Total number of work trips	0.534	2.005	2.091	2.043	2.504
Total number of non-work trips	2.459	4.721	6.168	5.041	4.324

Source Computed by authors

(i) Birth year of the household head (before 1945, 1946–1955, 1956–1965, 1966–1975 and after 1976); and
(ii) Household location (55 planning areas aggregated into 13 regions).

Table 2 shows the variation of the three travel demand measures with different birth year groups. Several patterns can be observed for the households' travel demand. Firstly, both the total number of trips and the number of non-work trips are highest for the households whose head was born between 1956 and 1965, and decrease gradually with heads of a younger or older generation. Moreover, the youngest generation (i.e., the birth year group born after 1976) has the highest work trip frequency. Further, the generation of trips for both work and non-work purposes is the lowest for the oldest generation. This is not surprising considering the potential inconvenience older people might have when they move. Lastly, for all birth year groups, non-work trip generation is much higher than work trip generation, which is consistent with the general trend that non-work trips constitute a large proportion of urban trips (Rajamani et al. 2003).

To test the effectiveness of the grouping criteria, we compared the between-group and within-group standard deviations of the key variables (Tsai et al. 2014). For all variables, the between-group standardised deviations are larger than the within-group standardised deviations, suggesting sufficient inter-group heterogeneity among cohorts. Hence, we concluded that the grouping criteria have created sufficient between-group variation and thus can be considered reliable. Overall, we obtained 195 (5*13*3) observations with an average of 106 households in each cohort.

5 Modelling Approach

The main concern of panel/pseudo panel estimation is that the panel units are likely to contain unobserved heterogeneity that is correlated with the error term. If this is the case, the pooled Ordinary Least Square (OLS) estimator will lead to biased and inconsistent estimations due to the violation of the error term independence. To control for unobserved heterogeneity, various other estimators are developed for panel/pseudo panel data models, among which the two most commonly used by researchers are the Fixed Effect (FE) and Random Effect (RE) estimators (Tsai et al. 2014). These two estimators differ in that the FE allows the unobserved heterogeneity to be correlated with the observed variables while the RE does not. The unobserved heterogeneity, in the case of this study, could be any characteristics that are time-invariant but potentially associated with the observed explanatory variables, such as travel preferences and attitudes towards the environment. For example, the travel preferences of households are likely to be related to their socioeconomic status. Households with higher income are generally more educated and thus tend to be more environmentally friendly, and deliberately choose public transport more often. Considering the potential association between the observed variables and the unobserved heterogeneity, FE estimator was utilised in the current study. In this way, the self-selection problem can be partially solved even though travel preferences are not directly measured. The corresponding equation is given below:

$$y_{it} = \boldsymbol{\beta}_1 \boldsymbol{SE}_{it} + \boldsymbol{\beta}_2 \boldsymbol{TS}_{it} + \boldsymbol{\beta}_3 \boldsymbol{BE}_{it} + \boldsymbol{\beta}_4 \delta_t + \alpha_i + \varepsilon_{it}$$

where y_{it} is the travel-related variable of interest for observation i at time t; $\boldsymbol{SE}_{it}$, $\boldsymbol{TS}_{it}$, and $\boldsymbol{BE}_{it}$ represent the vector of socioeconomic demographics, transport supply and the built environment for observation i at time t, respectively; $\boldsymbol{\beta}$ is the corresponding vector of coefficients; δ_t is time dummies, which is incorporated to capture the influence of aggregate trends over time; α_i denotes all other unobserved and time invariant factors for observation i that might affect household travel demand and travel mode choice; and ε_{it} is the random disturbance of observation i in time t. The fixed effect model assumes that $\boldsymbol{SE}_{it}$, $\boldsymbol{TS}_{it}$, and $\boldsymbol{BE}_{it}$ are correlated with both α_i and ε_{it}.

In the FE estimation, α_i is eliminated through a demeaned transformation to ensure the error term is not correlated with the repressors (Wooldridge 2013). One drawback of the FE estimator is that no time-invariant explanatory variables can be introduced in the model. However, since we are interested in unbiased and consistent coefficients of other variables, which are time-variant, the model fits the need of the study.

6 Estimation Results

Table 3 shows the estimation results of the FE models on overall trip frequency as well as work and non-work trip frequency. As expected, socioeconomic and demographic characteristics affect household travel demand significantly. More specifically, the age of the household head is positively related to the total number of trips and the total number of work and non-work trips. This is consistent with our previous findings that the age of the household head, a representative of the household generation, is a powerful indicator of household travel demand. Household size is also positively related with household work and non-work trip generation, which is to be expected since a larger family tends to have greater out-of-home travel needs, such as grocery shopping or family leisure activities. A 1% increase in household size translates to an average of 0.97%, 0.44% and 1.13% increase in overall trip frequency, work trip frequency and non-work trip frequency, respectively. In this regard, household size affects travel demand for non-work purposes more significantly than that for work purposes. Counter-intuitively, household income level does not show a statistically significant influence on the number of non-work trips, but it does show a positive impact on the overall number of trips and the number of work trips.

Table 3 Fixed effects models on total number of trips, work trips, and non-work trips

Explanatory variables	Total number of trips		Total number of work trips		Total number of non-work trips	
	Coef.	t-stat	Coef.	t-stat	Coef.	t-stat
Age of head of household	0.368**	4.21	0.750**	3.65	0.432**	3.16
Household size	0.965**	5.83	0.437**	2.49	1.130**	5.84
Household income	0.208**	2.33	0.769**	10.87	0.061	0.59
Car ownership	−0.173	−0.67	−0.531*	−2.03	−0.026	−0.08
Distance to the nearest expressway exit	0.133	1.49	0.019	0.26	0.165	1.51
Distance to the nearest LRT/MRT station	0.045	0.12	0.346	0.69	0.077	0.16
Walking path density	0.174	0.91	0.038	0.18	0.298	1.26
Land use entropy index	0.154**	2.29	0.112	1.37	0.163*	1.90
Population density	0.327**	2.22	0.481**	2.63	0.279*	1.76
Yr2004	−0.164	−1.45	−0.221*	−1.96	−0.154	−1.16
Yr2008	0.046*	1.80	−0.005	−0.18	0.063**	2.07
R-squared		0.9263		0.9691		0.8819
Adj R-squared		0.8799		0.9496		0.8074
Root MSE		0.1224		0.1258		0.1445

Note: ** Significant at $\alpha = 0.05$; * Significant at $\alpha = 0.1$

According to the results, household car ownership affects work-trip-making negatively, suggesting that the more cars a household owns, the fewer work trips household members will make. Note that trip frequency in this study was calculated at the household level. The negative relationship between car ownership and the number of work trips might be explained by a possibility that, for households with more than one cars, some household members are not routine workers and do not go to work everyday. Future research on household travel demand, which accounts for household structure, will shed more light on this subject.

With regard to transport supply, both rail transit accessibility and expressway proximity have no significant impact on household travel demand, suggesting that rail transit construction and expressway expansion may not stimulate or suppress the travel desire of households in the context of Singapore. However, this does not necessarily mean that transport supply has zero influence on household travel from other aspects. Previous studies have suggested that improving transit accessibility through rail transit construction can encourage the ridership of rail transit (Ewing et al. 2013), while expressway expansion tends to promote additional vehicle usage (Cervero 2003; González and Marrero 2012). The influence of transport supply on household travel mode choices merits further investigation in the future.

Among the built environment variables, only population density is positively associated with all the three travel demand indicators. A 1% increase in the population density leads to an average of 0.33% increase in total trip frequency, 0.48% increase in work trip frequency, and 0.28% increase in non-work trip frequency. Moreover, non-work trip generation tends to increase if the land is put to more diverse uses, such as residential, commercial, cultural or entertainment uses. The observed effect suggests that an increase in potential opportunities for activities to undertake around residences might induce additional non-work activities. However, land use mix does not have a significant influence on the frequency of work trips. Further, walking path density is not statistically significant in none of the three models, indicating that the walking environment does little to explain household work or non-work travel demand in Singapore. However, it is important to note that our measures of the built environment are computed as the subzone level. As subzones are aggregated enumeration units, some effects of the built environment may be obscured.

It is interesting to note that the coefficients on household travel demand of socioeconomic and demographic characteristics are relatively larger in magnitude than those of the spatial attributes (land use mix in particular). Moreover, most of the built environment variables are not statistically significant in explaining household travel generation. In this regard, household travel demand, either for work or non-work purposes, is more strongly influenced by the households' inherent attributes (socioeconomic and demographic characteristics), while external physical factors, such as transport supply or the living environment, have a less impact. These findings in the context of Singapore echo those from previous research based in the American or European cities, which indicates that trip frequency is primarily determined by travellers' socioeconomic characteristics and less associated with the built environment (Ewing and Cervero 2010; Ewing et al. 1996; Zhang 2004). Hence, land use policies may play a less critical role in controlling people's travel demand than often assumed.

7 Conclusions

This paper has investigated the influence of transport supply on household travel demand in the context of Singapore. Specifically, it examines the effect of improved transport accessibility induced by rail transit construction and expressway expansion on work and non-work travel frequency of households. Two questions have been addressed: does transport infrastructure investment increase or suppress household travel demand; and if so, how and to what extent transport infrastructure investment influences household travel demand. Travel demand is represented by the total number of trips made by a household on one random weekday. Due to the lack of longitudinal data, this study has adopted a pseudo panel approach, constructing a pseudo panel dataset with household travel surveys conducted at three different years (2004, 2008 and 2012). Separate models have been developed to estimate work and non-work trips considering that the decision-making process tends to differ for trips with different purposes.

We find that for both work and non-work purposes, rail transit accessibility and expressway proximity have no considerable impact on household travel demand. The result suggests that improved accessibility brought by rail transit construction and expressway expansion may not stimulate or suppress the travel desire of Singaporeans. Thus, induced travel demand by expressway expansion is not a critical issue in Singapore. However, the potential positive impact of rail transit provision, such as more usage of public transport instead of cars, merits further consideration in transport planning in Singapore.

We also find that households living in neighbourhoods with higher density tend to travel more. Land use diversity affects only non-work trip generation. In this regard, planners can consider mixed land use strategies by providing diverse opportunities (e.g., recreation, retails, exercise and employment) around residential areas in order to encourage more out-of-home trips, and thus create opportunities for people to meet and interact with each other. As Singapore society ages rapidly, this is also important for older adults who may not be able to travel far but can still participate in various activities near their residence. In the New Urban Agenda[6] which was adopted in Quito in October 2016, spatial organisation, accessibility and design of the space are recognised as an important determinant of people's travel patterns and thus could help promote or hinder social cohesion and inclusion. Appropriate land use planning therefore has the potential to contribute to the achievement of the

[6]For more information on the New Urban Agenda, see http://habitat3.org/wp-content/uploads/NUA-English.pdf.

Sustainable Development Goals[7] (SDGs) and targets, especially SDG 11 of making cities and neighbourhoods inclusive, resilient and sustainable.

In general, our findings are consistent with those found elsewhere, where household travel demand, either for work or non-work purposes, is more strongly influenced by households' socioeconomic and demographic characteristics, while the surrounding living environment plays a less important role. Hence, urban planning policies may be less effective in controlling people's travel demand than they are often assumed to be. However, this does not necessarily mean that planning policies are not effective in managing travel behaviour at all. Numerous studies have shown that a properly-designed neighbourhood, for example, with enhanced walkability, will encourage people to walk outdoors more and, therefore, get more physical activity; this enhances the 'sense of belonging' to a place due to greater interaction among neighbours, which contributes to neighbourhood or city cohesion and resilience.

The study can be extended further in several directions. First, even though a pseudo panel approach has been able to compensate for the lack of real panel data in many ways, the impact of transport supply on travel demand can be better understood if genuine panels at the individual level are available. Second, in this research, the residential self-selection issue is controlled by incorporating cohort fixed effect. In future research, informed psychological variables, such as travel attitudes, could be incorporated explicitly to better address the residential self-selection issue. Third, the current study considers only travel demand. Future research may investigate other aspects of travel behaviour, such as transport mode, vehicle usage and car ownership, that are also relevant to policymaking.

Acknowledgements This research is supported by the PhD research scholarship, National University of Singapore (NUS) and the Musim Mas Dissertation Award for Research on Sustainability, Business School, NUS. The authors appreciate the support of the Future Urban Mobility (FM), Singapore-MIT Alliance Research and Technology (SMART) Centre, and the Singapore Land Transport Authority for providing the data used in this study.

References

Bernard J-T, Bolduc D, Yameogo N-D (2011) A pseudo-panel data model of household electricity demand. Res Energy Economics 33(1):315–325. https://doi.org/10.1016/j.reseneeco.2010.07.002

Boarnet MG, Nesamani KS, Smith S (2003) Comparing the influence of land use on nonwork trip generation and vehicle distance traveled: an analysis using travel diary data. Paper presented at the the 83rd annual meeting of the transportation research board, Washington, DC

Brownstone D (2008) Key relationships between the built environment and VMT. In: Transportation research board special report 298: Driving and the built environment: effects of compact development on motorized travel, energy use, and CO_2 emissions. Washington, DC: National Academies Press

[7] For more information on the Sustainable Development Goals, see https://sustainabledevelopment.un.org/?menu=1300.

Brownstone D, Golob TF (2009) The impact of residential density on vehicle usage and energy consumption. J Urban Economics 65(1):91–98. https://doi.org/10.1016/j.jue.2008.09.002
Cervero R (2003) Road expansion, urban growth, and induced travel: a path analysis. J Am Plann Association 69(2):145–163. https://doi.org/10.1080/01944360308976303
Crane R (1996). Cars and drivers in the new suburbs. American Planning Association. J Am Plann Association 62(1):51. https://doi.org/10.1080/01944369608975670
Dargay JM (2007) The effect of prices and income on car travel in the UK. Transp Res Part A 41(10):949–960. https://doi.org/10.1016/j.tra.2007.05.005
Dargay JM (2001) The effect of income on car ownership: evidence of asymmetry. Transp Res Part A Policy Practice 35(9):807–821. https://doi.org/10.1016/s0965-8564(00)00018-5
Dargay JM (2002) Determinants of car ownership in rural and urban areas: a pseudo-panel analysis. Transp Res Part E 38(5):351–366. https://doi.org/10.1016/s1366-5545(01)00019-9
Deaton A (1985) Panel data from time series of cross-sections. J Econometrics 30(1):109–126. https://doi.org/10.1016/0304-4076(85)90134-4
Elldér E (2014) Residential location and daily travel distances: the influence of trip purpose. J Transp Geography 34:121–130. https://doi.org/10.1016/j.jtrangeo.2013.11.008
ESRI (2010) Network analyst tutorial. In. Retrieved from http://help.arcgis.com/en/arcgisdesktop/10.0/pdf/network-analyst-tutorial.pdf
Ewing R, Cervero R (2001) Travel and the built environment: a synthesis. Transp Res Record J Transp Res Board 1780(1):87–114. https://doi.org/10.3141/1780-10
Ewing R, Cervero R (2010) Travel and the built environment: a meta-analysis. J Am Plann Association 76(3):265–294. https://doi.org/10.1080/01944361003766766
Ewing R, DeAnna M, Li S-C (1996) Land use impacts on trip generation rates. Transp Res Record J Transp Res Board 1518(1):1–6. https://doi.org/10.3141/1518-01
Ewing R, Hamidi S, Gallivan F, Nelson A, Grace J (2013) Combined effects of compact development, transportation investments, and road user pricing on vehicle miles traveled in urbanized areas. Transp Res Record J Transp Res Board 2397:117–124. https://doi.org/10.3141/2397-14
Frank L, Kerr J, Chapman J, Sallis J (2007) Urban form relationships with walk trip frequency and distance among youth. Am J Health Promotion, 21(4_suppl):305–311. https://doi.org/10.4278/0890-1171-21.4s.305
González RM, Marrero GA (2012) Induced road traffic in Spanish regions: a dynamic panel data model. Transp Res Part A Policy Practice 46(3):435–445. https://doi.org/10.1016/j.tra.2011.11.003
Handy S, Cao X, Mokhtarian P (2005) Correlation or causality between the built environment and travel behaviour? Evidence from Northern California. Transp Res Part D Transp Environ 10(6):427–444. https://doi.org/10.1016/j.trd.2005.05.002
Kitamura R (2009) A dynamic model system of household car ownership, trip generation, and modal split: model development and simulation experiment. Transportation 36(6):711–732. https://doi.org/10.1007/s11116-009-9241-9
Kitamura R, Mokhtarian PL, Laidet L (1997) A micro-analysis of land use and travel in five neighborhoods in the San Francisco Bay Area. Transportation 24(2):125–158. https://doi.org/10.1023/a:1017959825565
La Paix Puello LC, Monzón A, Cherchi E, Loro M. (2013) Residential self-selection and geographical scales in trip frequency. Paper presented at the 13th world conference on transportation research, Rio de Janeiro, Brazil
Meng Y, Brennan A, Purshouse R, Hill-McManus D, Angus C, Holmes J, Meier PS (2014) Estimation of own and cross price elasticities of alcohol demand in the UK—A pseudo-panel approach using the living costs and food survey 2001–2009. J Health Economics 34:96–103. https://doi.org/10.1016/j.jhealeco.2013.12.006
Milakis D, Cervero R, van Wee B (2015) Stay local or go regional? urban form effects on vehicle use at different spatial scales: a theoretical concept and its application to the San Francisco Bay Area. J Transp Land use 8(2). https://doi.org/10.5198/jtlu.2015.557

Mokhtarian PL, Cao X (2008) Examining the impacts of residential self-selection on travel behavior: a focus on methodologies. Transp Res Part B Methodol 42(3):204–228. https://doi.org/10.1016/j.trb.2007.07.006

Rajamani J, Bhat CR, Handy S, Knaap G, Song Y (2003) Assessing impact of urban form measures on nonwork trip mode choice after controlling for demographic and level-of-service effects. Transp Res Record J Transp Res Board 1831(1):158–165. https://doi.org/10.3141/1831-18

Scheiner J, Holz-Rau C (2013) Changes in travel mode use after residential relocation: a contribution to mobility biographies. Transportation 40(2):431–458. https://doi.org/10.1007/s11116-012-9417-6

Shay E, Khattak AJ (2012) Household travel decision chains: residential environment, automobile ownership, trips and mode choice. Int J Sustainable Transp 6(2):88–110. https://doi.org/10.1080/15568318.2011.560363

Targa F, Clifton KJ (2005) The built environment and trip generation for non-motorized travel. Paper presented at the National Household Travel Survey Conference: Data for Understanding Our Nation's travel, Washington, DC

Troncoso Parady G, Chikaraishi M, Takami K, Harata N (2014) A panel data approach to understanding the effect of the built environment on travel behavior. Urban Regional Plann Rev 1:18–38. https://doi.org/10.14398/urpr.1.18

Tsai C-H, Leong W, Mulley C, Clifton G (2013) Examining estimation bias and efficiency for pseudo-panel data in travel demand analysis. Transp Res Record J Transp Res Board 2354:1–8. https://doi.org/10.3141/2354-01

Tsai C-H, Mulley C, Clifton G (2014) A review of pseudo panel data approach in estimating short-run and long-run public transport demand elasticities. Transp Rev 34(1):102–121. https://doi.org/10.1080/01441647.2013.875079

Verbeek M, Nijman T (1992) Can cohort data be treated as genuine panel data? Empirical Economics 17(1):9–23. https://doi.org/10.1007/bf01192471

Weis C, Axhausen KW (2009) Induced travel demand: Evidence from a pseudo panel data based structural equations model. Res Transp Economics 25(1):8–18. https://doi.org/10.1016/j.retrec.2009.08.007

Wooldridge JM (2013) Advanced panel data methods. In: Wooldridge JM (ed) Introductory econometrics: a modern approach, 5th edn. South-Western Cengage Learning, Mason, Ohio, pp 484–511

Zhang M (2004) The role of land use in travel mode choice: evidence from Boston and Hong Kong. J Am Plann Assoc 70(3):344–360. https://doi.org/10.1080/01944360408976383

Zhu Y, Diao M (2016) The impacts of urban mass rapid transit lines on the density and mobility of high-income households: a case study of Singapore. Transport Policy. https://doi.org/10.1016/j.tranpol.2016.03.013

Participatory Design to Co-create Community Spaces

Im Sik Cho and K. C. Ho

Abstract The extensiveness of public housing in Singapore, where more than 80% of its residential population lives, underscores the importance of the design and provision of shared amenities and communal spaces in shaping social bonding. Building social resilience in neighbourhoods has to do with creating opportunities for residents to interact on the assumption that residents who are socially connected in their immediate neighbourhood are also more likely to be involved in helping others and in contributing to the well-being of the neighbourhood. This chapter discusses a research project that aims to establish a participatory design approach to co-create community spaces in an existing neighbourhood. It is expected that the outcomes of this study will reveal key challenges in implementing participatory planning in local neighbourhood context and offer insights into potential avenues for strengthening community's capacity to initiate collective action, which is essential to support long term social sustainability and resilience of neighbourhoods.

Keywords Singapore · Public housing · Neighbourhood · Community

1 Introduction

Since the 1960s, when the state launched its public housing programme, public housing has played an important role in the development of communities in Singapore, as it implies far more than the mere provision of housing space. The extensiveness of public housing throughout Singapore—where more than 80% of its residential population lives, develops social relationships, and shares common experiences—sets

I. S. Cho (✉)
Department of Architecture, National University of Singapore, Singapore, Singapore
e-mail: akicis@nus.edu.sg

K. C. Ho
Department of Sociology and Asia Research Institute, National University of Singapore, Singapore, Singapore
e-mail: sochokc@nus.edu.sg

C.-H. Leong and L.-C. Malone-Lee (eds.), *Building Resilient Neighbourhoods in Singapore*, Advances in 21st Century Human Settlements,
https://doi.org/10.1007/978-981-13-7048-9_6

the foundation for the use of public housing for social cohesion as a conscious goal of the programme in the city-state.

Until recently, however, Singapore's planning process has been largely led by the state to realise strategic goals that sought little participation by the private sector, much less by individuals (Hee and Ooi 2003). While planning efforts have traditionally taken a highly centralised approach, the main challenge has been in incorporating public participation, which is often regarded as crucial by many in the planning circles (Sanoff 2000; Innes and Booher 2004; Irvin and Stansbury 2004; Ooi 2004; Baker et al. 2006), and in shifting cultural expectations from central to participatory planning.

Participatory planning advocates have argued that participation in local neighbourhoods can enhance social capital, social cohesion, local ownership and sense of belonging among neighbours (Putnam 2000) and higher levels of attachment to their neighbourhoods (Taylor 1996). These findings support greater community-based participatory processes in the design and development of neighbourhoods. Public participation in urban programming initiatives is not only a key element towards local sustainability (Holden 2011) but is also regarded as an effective way to build social resilience, which in turn helps to build overall urban resilience (Jha et al. 2013).

In recent years, Singapore's changing social composition and political transformation has increased interest from both the government and grassroots groups to engage the local community and institute more community involvement at the local level (Cho 2017; Cho and Križnik 2017). In this context, this chapter will discuss a recent research project, which tested a 'three-pronged' approach to community building in an existing public housing neighbourhood in Singapore, encompassing attention to not only the 'physical design' but also its 'programmatic' and 'organisational' aspects. The project has attempted to facilitate active participation of the community and various stakeholders in the planning and design of their neighbourhood spaces in order to encourage a sense of ownership and create stronger bonds among members of the community. A participatory design approach whereby the various stakeholders are included throughout the process of developing the proposed design typologies was adopted for co-creating with the residents two community space typologies named "Neighbourhood Incubator" and "Social Linkway" with interventions at various points along a walkway that runs through the neighbourhood (Cho 2016; Cho and Križnik 2017). By illustrating the processes and challenges in testing out the various methods to engage the community, this chapter offers insights into potential avenues for strengthening community's capacity to co-manage common spaces and initiate change in the neighbourhood through collective action, which is essential to support long term social sustainability and resilience of neighbourhoods.

This chapter has two objectives. The first objective is to examine the nature of participation among residents in the creation of new amenities for the Tampines Central neighbourhood. The second is to highlight the way such amenities have improved the interaction among the residents through their use of the amenities.

The chapter is organised in six sections. After *Introduction* (Sect. 1), *Research Background* (Sect. 2) provides a brief overview of Phase 1 and Phase 2 of the research. Section 3 *Establishing a Participatory Design Approach to Co-create Community*

Spaces illustrates the four main phases for participatory planning and community engagement established by the research. Section 4 *Measuring Everyday Consequences of Participatory Approach on Neighbourhood Communal Life* focuses on the analysis from the study, followed by Sect. 5 that discusses the potential and challenges for building resilient neighbourhoods and *Conclusion* (Sect. 6).

2 Research Background

In 2012, the National University of Singapore (NUS) and the Housing and Development Board (HDB) embarked on a collaborative research project to study the impact of the built environment on community bonding in Singapore's public housing neighbourhoods. Ultimately, the research shed light on how existing facilities and amenities are used by residents, how effective they have been in fostering social interactions amongst neighbours, and the current state of neighbourhood bonding in Singapore. The study has contributed to identify designs that can facilitate better community interaction, as the survey results suggest that residents who frequently use a wide range of amenities in their precinct tend to report a stronger sense of place and belonging for their precinct. Nine design strategies and six typologies of community spaces (i.e., playground clusters, flexible open spaces, neighbourhood incubators, social linkways, shared verandas, and green social decks) were developed from this investigation and a follow-up study (or second-phase) was conducted during 2014–2015 to implement the recommended design typologies in an existing public housing neighbourhood in Singapore (Centre for Sustainable Asian Cities 2014, 2015; Cho 2016; Cho and Kržnik 2017).

The first phase of the research (2012–2014) argued that a more holistic approach to planning and design would reflect not only aspects of design infrastructure, programmes, and activities but also the level of organisational support. For example, the sustainability of community-led projects highly depends on the level of organisational support, which often drives and facilitates both physical and programmatic aspects of a neighbourhood project. Central to establishing a participatory approach to neighbourhood planning and design are volunteerism and grassroots structures that involve a variety of stakeholders. Moving beyond the existing engagement practices in Singapore, which are broadly consultative in nature, to identify, implement and adapt strategies with an aim to establish stronger and well-structured institutional support systems can ensure sustainability and demonstrate social impact and capacity building. Such strategies include creating a platform for collaborative planning where various stakeholders and residents can work together to improve their neighbourhoods or introducing a neutral facilitator and supporting organisations to implement effective and meaningful participatory processes (Centre for Sustainable Asian Cities 2014, 2015; Cho 2016, 2017; Cho and Kržnik 2017; Cho et al. 2017).

The second phase of the research study (2014–2015) aimed to examine and implement possible mechanisms and platforms of participation, in order to encourage deeper social interaction and heighten the sense of community and belonging to a place. It aimed to propose a potential participatory mechanism suitable for the context of Singapore's neighbourhood planning and to engage key stakeholders in developing the mechanism through an intensive participatory process.

This chapter will focus on the second-phase study that aimed to further develop and improve the proposed design guidelines through a mechanism of community design and participation.

The research was conducted in the form of "action research" whereby the researchers were actively engaged in the community design participation process. Data collection was done throughout the participation process, whereby every data gathered from each action or step was reflected back to the whole research process for the improvement of the designed mechanism (Centre for Sustainable Asian Cities 2015). More precisely, this research employed the Participatory Action Research (PAR) method, which emphasises participation and action for research in communities. With this approach, a platform for collaboration among different stakeholders involved or affected by the research study was created, in which the community members were considered as co-researchers to address questions and issues pertaining to this project.

The engagement process with residents involved the use of pop-up booths along commonly visited areas. The booths served as a way to inform and involve residents to participate in the planning of their neighbourhood in terms of amenities which they will like to see built. This was followed by a focus group discussion to collect ideas on neighbourhood improvements, a design workshop where these ideas were further elaborated and developed into scale models which can be constructed. The fourth and final event was a co-creation art event where residents were invited to paint, guided by artists who were brought into help facilitate the process (Centre for Sustainable Asian Cities 2015; Cho 2016; Cho and Križnik 2017). More than 130 surveys have been completed and subsequent qualitative analysis was conducted by a group of sociologists to reveal more insights on the success of the engagement process, residents' satisfaction and the social implications of the usage of implemented public amenities.

3 Establishing a Participatory Design Approach to Co-create Community Spaces

Through an intensive community engagement process, the research project aimed to establish a model of participatory planning and community engagement, which can work in the local neighbourhood context, consisting of four main phases described below.

3.1 Understand the Local Context of Participatory Planning

It is crucial to start with understanding the local context, in order (i) to assure that the project responds to the actual local community needs as well as to capitalise on existing neighbourhood resources and opportunities, and (ii) to ensure the success of any participatory process. Forming relationships with neighbours is important in building communities and in encouraging more participation, which is a challenge because, according to the interview key findings from the first phase of the research (2012–2014), many do not even know their neighbours (Centre for Sustainable Asian Cities 2014). Responding to local context connotes three important activities: (i) understanding the community's demographic, cultural, socio-economic and political influences[1]; (ii) establishing a common interest; and (iii) involving diverse views in the collaborative process.

The project has reached out to the residents living in a public housing neighbourhood in Tampines Central that consisted of 5000 household units, specifically those in block numbers 830–863 at Tampines Street 82, 83, and Avenue 5 in Singapore. The research team conducted several site visits in July to August 2014 to observe and experience the daily life of the residents, map their meeting places, build relationships with the local connectors, and locate possible sites for planning new neighbourhood amenities. To better understand the current level of community participation in the neighbourhood context, formal interviews were conducted in August 2014 with various local stakeholders, such as leaders and members of grassroots organisations and representatives from the HDB and the town council.

3.2 Collaborate with Community Stakeholders

Community participation involves various stakeholders, who are commonly defined as "any individual or group who can affect or is affected by the actions, decisions, policies, practices, or goals of the organisation" (Carroll 1996, p. 74; Baker et al. 2006 cited in Buchholz and Rosenthal 2004, p. 144). Allen and Cars (2001) have emphasised that in "multicultural environments", there is a greater need for "bottom-up government structures" and more "flexibility to evolve over time".

In the workshop with stakeholders held in August 2014 at Tampines Central, various issues were discussed, including (i) each stakeholder's definition of successful community participation; (ii) their main takeaways from this research study; (iii) the community's assets and issues; and (iv) the role and contribution of each stakeholder in the project. In addition to the HDB and the local community in Tampines, the research team collaborated closely with local stakeholders, including the Tampines town council and the chairpersons of the three Residents' Committees (RCs) within

[1]The effectiveness of any public engagement approach is highly dependent upon "the political and cultural milieus of a country and the readiness and maturity of its civil society to propose ideas and responsibly see them through" (Hee and Khoo 2013).

the project site (Tampines Ville, Tampines Parkview and Tampines Palmwalk RCs). The People's Association's Tampines Central constituency office was also involved in the project as a member of the working committee (Cho 2016). Involving local stakeholders throughout the process is crucial as they not only can point out long-term implementation and operational concerns, but it also empowers them to co-create and sustain initiatives for the community over the long run.

3.3 Engage with Community

A balanced combination of formal and informal forms of participation and feedback collections were used, with formal methods such as Focus Group Discussions (FGDs), and informal channels, such as newsletters/online websites, interactive boards, pop-up events, and volunteers' programme. The well-suited mix of various engagement methods ensured enhanced participation from the residents in terms of number and diversity, and improved the current mainstream practice where feedback is informally collected over a period of time to make decisions upon which neighbourhood amenities are upgraded.

Several *community outreach events involving pop-up booths* (Fig. 1) with exit questionnaires were organised during September–October 2014, which aimed at reaching out to a larger community who might not participate in the FGDs and design workshops. Three pop-up events were held to display the project and its progress and seek residents' feedback. These pop-up events were held in public spaces such as void decks, green open spaces, and hard courts within the neighbourhood of Tampines Central. The pop-up events happened at different times of the day on both weekdays and weekends to tap on different pedestrian flow as well as residents of different demographic groups in order to reach as many people as possible. Participants who came to the pop-up booths were engaged to identify their favourite places in the neighbourhood by placing dot stickers onto a map and describing the reasons for their choices on stickers provided by the team.

As *co-learning with communities* is important, one FGD with residents was conducted in September 2014 to further understand the neighbourhood as well as its residents' daily activities, gathering spots, nodes, landmarks, and their thoughts on the quality of the current neighbourhood amenities. Residents shared the types of locations where they often meet their neighbours. The FGD also gathered residents' ideas for constructing the proposed design typologies (i.e., the Neighbourhood Incubator and the Social Linkway) and the locations for their implementation.

Following the FGD, a *co-design workshop* (Fig. 2) was conducted in October 2014 to engage the residents in visualising their ideas and aspirations for constructing the proposed design typologies. Potential subjects were recruited through posters that the residents' committees distributed via their notice boards and websites, grassroots leaders, and during the community outreach events/pop-up booths. Despite the initial concerns of the research team regarding the residents' lack of spatial understanding, the design workshop was conducted successfully, perhaps due to the

Fig. 1 Engaging the residents (Pop-up event). *Source* National University of Singapore; Housing and Development Board, Singapore

residents' familiarity with their environment and the facilitators' encouragement to the participants to share diverse and interesting ideas. The residents were willing to elaborate their suggestions and feedback during the session, and some residents expressed their interest to get involved even deeper in the project with more commitment, if they had more time.

Finally, *co-creation activities* were organised in May 2015 aiming to involve the residents in the actual construction of the two design typologies. During the co-creation events, residents participated in activities such as painting and creating community arts with artists. The art-making events were successful in getting people involved and allowing them to play a role in the overall process as some residents took pride in their efforts and contribution. These events were successful as many residents came back over several days to participate and contribute further. However, some residents, especially adults, tended to step back and observe rather than participate. Perhaps because it is a painting activity done in a public area, the adults were more reserved and remained observers, preferring to let their children do the painting. The younger kids were already used to painting in their kindergartens and schools and enjoyed the activity. Nevertheless, the co-creation events helped to generate greater awareness of the project and made residents feel that they had a stake in the construction of the amenities, as well as a 'sense of ownership' of the built environment, which is crucial for the long term sustainability of the project.

Fig. 2 Engaging the residents (Design workshop). *Source* National University of Singapore; Housing and Development Board, Singapore

4 Measuring Everyday Consequences of Participatory Approach on Neighbourhood Communal Life

This research project adopted a participatory approach to co-create community spaces in a neighbourhood context, beginning with community involvement at the planning stage and culminating in the construction of new community spaces. However, it is hoped that not only community spaces but also a stronger 'sense of community' was built in the process. In order to measure this outcome, a group of sociologists have been involved to observe whether there are any changes in the interaction pattern among the residents and in the usage of community spaces after the construction of the design typologies, through various research tools, including video analytics.[2] Besides observation, structured questionnaires with the participants of the community events and interviews with the various stakeholders were also carried out to measure the change in their sense of community caused by participating in the process.

[2]The data collection procedures were approved after review by the NUS Institutional Review Board (IRB). IRB approval NUS2330, NUS-IRB reference code A-14-154, approval of protocol amendment 3 dated 17 April 2015.

4.1 Resident Participation

The research team administered and collected questionnaires from four events. First, the pop-up booths in four different places in September 2014 in the research site designed to create awareness of which 389 residents participated; this is the simplest form of involvement. Second, the following week in September 2014, a FGD was held at Junyuan Secondary School on a Saturday. This event brought together 94 residents, who were divided into groups and discussed practical ways of improving their neighbourhood. Third, a design workshop was held in October 2014 to collect ideas and feedback on problems, in which 43 residents participated. Fourth, in May 2015, the research team organised a further set of co-creation events, which encouraged the residents to participate in two community art projects where 130 participants contributed their artistic ideas.

Not all residents turned up for every event. And not every resident who turned up filled in the questionnaires. For example, of the 300 residents who participated in the pop-up booths, 94 filled in the survey. The FGD conducted on 27 September attracted 94 participants and only 26 participants filled in the questionnaire mainly because the FGD stretched over an hour and the participants were ready to go home. The design workshop on 11 October had 43 participants, and this time a more encouraging 32 participants filled in the questionnaire.

From the FDG and Design Workshop, we note that relatively few participants filled in the survey questionnaire. The small number means that the analysis of survey responses can only be indicative and not conclusive. If the participatory action research (PAR) perspective which was adopted in this research "recognises the need for persons being studied to participate in the design and conduct of all phases (e.g., design, execution, and dissemination) of any research that affects them" (Vollman et al. 2004 cited in MacDonald 2012, p. 20), then this very small number is indicative of the weak outcome of PAR in Singapore. Since the research team has tried to encourage the residents who have participated in previous events to turn up, their failure to do so suggests that PAR has some way to go before it becomes an established channel for participation. One lesson from this exercise is the need to work with residents for a longer period of time 'before' asking them to participate. Perhaps a period of up to three months can be worked into the planning schedule and within this period, scheduled meetings can allow the importance of participation to be discussed. This period is also important to allow residents to get to know each other. This longer period of involvement should take the form of preparing them for participation. This preparation should involve (i) getting residents to understand their role as residents, (ii) letting them interact with one another informally, and (iii) having an office in the town council or residents' committee's office or in the residents' committee's Facebook for them to express their initial views.

The analysis of the survey data was designed to answer two questions. The first question was to examine if those who participated in the organised events (the FGD, Design Workshop and Co-creation Event) are different (in terms of their orientation towards the neighbourhood) from the passersby who happen to be there during the times when the pop-up booths happen to be operating. Typical of large cities, relations between neighbours are minimal because most residents tend to spend most of their time at work or in school. With less time spent in the neighbourhood and weak neighbourly ties, it is reasonable to expect that most HDB residents are unlikely to step forward as participants for events which the research team had organised. Is it possible to understand the profile of those that do participate? Moreover, the question of the underlying need which propels participation is an issue to take note of in Singapore. It is known from the phase 1 survey that residents in Singapore are "very satisfied" with the amenities in their neighbourhood. PAR literature review suggests that the roots of PAR exist in poor neighbourhoods where empowerment through engagement in the meeting of collective needs of the community forms the logic and the dynamic underlying the PAR process. It is therefore understandable that when residential satisfaction with their neighbourhood is high, participation may not be as strong.

Figure 3 groups the residents who happen to be passing by and filled in a short questionnaire when the pop-up booth was in operation. These are categorised as passersby. The second group is comprised of residents who participated in one or more of the following activities which required some commitment at their end: the FGD, design workshop, and the co-creation event. These groups are labelled as participants. The analysis was aimed to see if the participants were different in their attitudes compared to passersby who did not attend subsequent events. The difference between the means of the two groups was observed where the significant differences were in the three statements on the left of Fig. 3. This suggests two orientations on the part of the participants. Firstly, they (the participant group) were more likely to think that their efforts would make their neighbourhood a better place. Secondly, they could see that their neighbourhood had room for improvements.

The second question is whether those, who were repeat group of participants, had their attitudes towards participation and the neighbourhood changed as a result of repeated participation. Among the seven components of the PAR process highlighted by Selenger (1997), one involves the transformative effects of participation. PAR "fosters greater awareness of individual's resources, skills and capacities for self-reliant development" (Selenger 1997). If this is indeed the case, then the analysis was aimed to see if there was any evidence that participation had a transformative effect of changing the participants' view of their skills and their responsibility towards their precinct. If there was a change, then this project might be viewed as a modest attempt at fostering a grassroots type of participation.

Figure 4 shows the changes in attitudes of participants by comparing the first time they completed the questionnaire after the first event, with the next time they handed in a completed questionnaire after the second event. It should be noted that the sample size is very small at only 15 participants and, therefore, the results should be at best a rough approximation of how outlooks of participants might change over the co-creation exercise. The largest change recorded was with the item "any improvements can be managed by the local authorities" (1.8 in time 2 vs. 3.1 in time 1). This can be read as an increased confidence in local authorities handling neighbourhood improvements (indicating residents who saw the process as generally

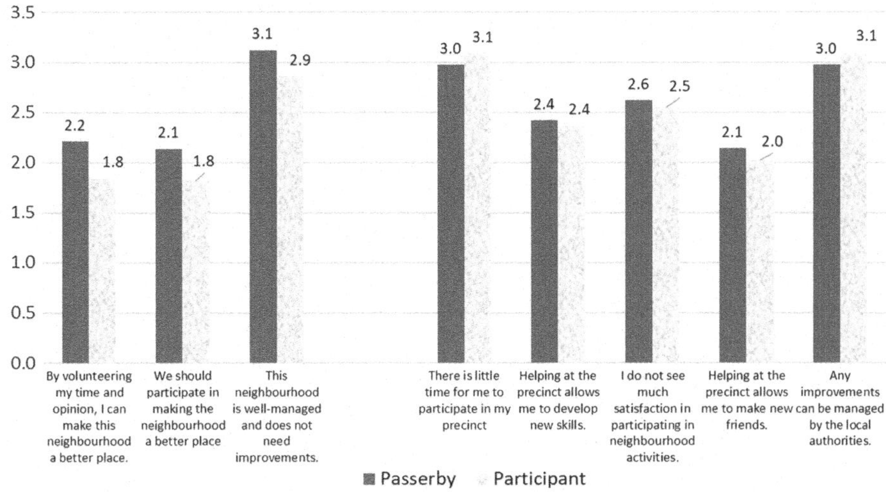

Fig. 3 Degree of participation in co-creation efforts between passersby and participants. *Note* 1 = Strongly Agree, 2 = agree, 3 = neutral, 4 = disagree, 5 = strongly disagree. *Source* Data obtained from Study on the Design Recommendations to Foster Community Bonding project

positive). It could also be read that perhaps after the participation process, which could be lengthy and frustrating when discussing a variety of conflicting ideas, participating residents had decided to hand the reins back to the local authorities (indicating residents saw the process as generally negative).

The highest scores for this repeat group of participants is recorded for the item "I don't see much satisfaction in participating in neighbourhood activities". The high mean scores 3.8 in time 1 and 4.0 in time 2, suggests that this small group of participants saw value in joining in neighbourhood activities. This group of repeat participants also responded well to the issue "there is little time for me to participate in my precinct". Having means of 3.2 in time 1 and 3.6 in time 2 suggests that they are even likely to disagree with the idea that they do not have enough time to participate. This is a positive development and suggests that social resilience may be sustained over time. However, the item "any improvements can be managed by local authorities" shows a dramatic increase in agreement between time 1 and time 2. This suggests a number of possibilities: Either they began to see that the authorities were doing a good work or the participants also realised the inherent difficulties of such projects and wanted to leave this type of difficult work to the authorities.

It can be argued that creating a better model for involving the participants in subsequent projects is essential, perhaps by (i) moderating their expectations, (ii) coaching them to accept the suggestions of other residents and not just to champion their ideas, and (iii) also keeping the consultation sessions to a manageable

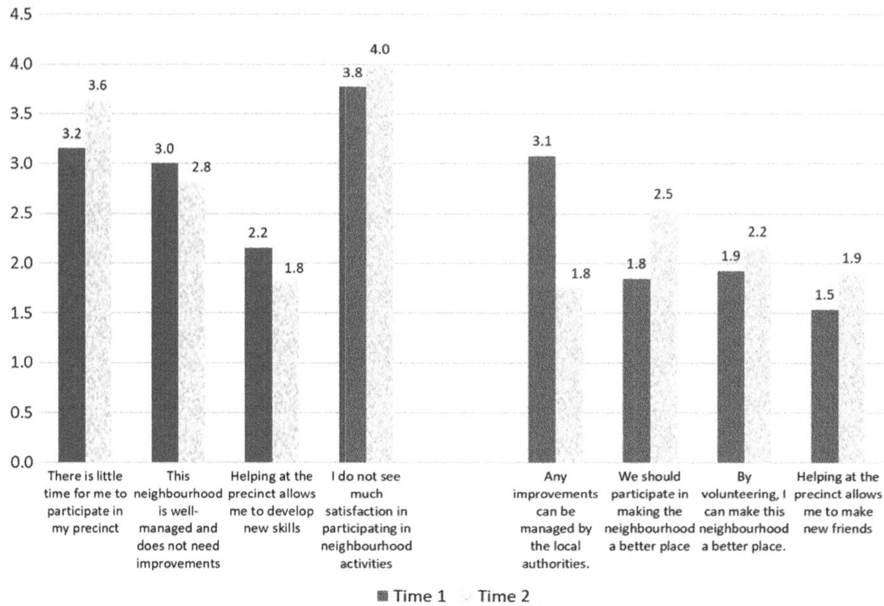

Fig. 4 Participants' change in outlook through subsequent participation. *Note* 1 = Strongly Agree, 2 = agree, 3 = neutral, 4 = disagree, 5 = strongly disagree. *Source* Data obtained from Study on the Design Recommendations to Foster Community Bonding project

time period.[3] And as the research project develops, it is important to find ways of sustaining their interest and maintaining the networks built over the consultation process.

4.2 Comparing Pre- and Post-intervention Traffic

Of the different projects suggested by residents, one of the most promising has been the building of the café at the void deck[4] of Block 839. Through CCTV cameras installed in this area, we were able to measure, during a three day period, the number

[3]Community engagement process took about 9–10 months in this research project (from the first workshop with community stakeholders held in August 2014 to the co-creation events held in May 2015). Our interview insights during phase 1 revealed that residents prefer an informal type of involvement rather than a formal structure, due to time constraints and level of commitment (CSAC 2015). This factor should be carefully considered during the research design to keep the engagement sessions manageable to the residents.

[4]A void deck is a term used to describe the first/ground floor of public housing blocks in Singapore. As the term implies, public housing residential buildings typically have the ground floor empty of residential units. Subsequently, this empty space may have different types of community activities (for example, childcare centre or senior citizen activity centre) added as the need arises. See https://en.wikipedia.org/wiki/Void_deck.

of users before and after the building of the café. Table 1 shows the number of users as well as the duration of usage of the void deck for comparable time periods (the morning period and Tuesdays, Friday and Sunday). It can be observed that after the Palmwalk Café (Learning Link within the "Social Linkway") was built and in operation, the number of users increased from a mean of 10 users for the table and chairs to a significant 55 users for the Palmwalk Café. Clearly, the offer of beverage and snacks in the morning brings residents to the void deck at Block 839. It is noted that there are regulars, such as the group of elderly men who sit, chat and play board games like chess or checkers at the void deck. It is also observed that single adults have a hot beverage before going off to work; people who finish their exercise class in the nearby play area also come to the café for a drink and snack. Before the café was built, the void deck only had a table and some seating. These served the purpose of a short period of resting for most residents before moving on with their journey. A smaller group of youths also made use of the place for study. Having a café serving refreshments and snacks at regular hours and having a regular group of patrons mean that there are incentives to stay in the place longer. As a result the average duration of usage of the void deck space doubled from 10.9 min to 21.6 min.

The Palmwalk Café (Learning Link within the "Social Linkway", Fig. 5), by all accounts was quite a success, generating between 20 and 30 users for a 3 and a half hour morning period. Residents find the spot a convenient place to chat while having a cup of coffee, members of the morning exercise group make a rest stop after their workout, and workers visit the place to have a quick cup of beverage before heading to work. There may have been some displacement when the concrete table and chairs were removed to make way for the new café space. Palmwalk Café's intervention resulted in a new regularity among residents, and the CCTV footage showed that a significant amount of social interaction among the users was observed.

Table 1 Pre- and post-intervention (Void Deck at Block 839, Tampines Palmwalk RC)

	Pre-intervention	Post-intervention
Average Number of Users in a three day period (Tuesday, Friday and Sunday)	10.3	54.6
Average Duration of Usage (in minutes) in a three day period (Tuesday, Friday and Sunday)	10.89	21.63

Note Time period each day was from 8 am to 11.30 am, except on Sunday 13 Dec which ended at 12.15 pm instead of 11.30 am

Source Data obtained from Study on the Design Recommendations to Foster Community Bonding project

Fig. 5 Palmwalk Café (Learning Link within the "Social Linkway"). *Source* Housing and Development Board, Singapore

5 Potential and Challenges for Building Resilient Neighbourhoods

In summary, resident participation figures, and the data presented in Figs. 3 and 4 and Table 1 indicate the following about participation and resilience. Participation rates among residents tend to be very low. This is to be expected in urban neighbourhoods as residents are busy, and neighbourly relations tend to be superficial. In spite of the low participation rates, Fig. 3 shows that the small group of residents who participated in the engagement process believed that their role was important in making their neighbourhood a better place. As participants got involved in different co-creation activities, Fig. 4 shows that participants saw even greater value in participation over time. This is an important indicator of resilience as sustained participation for this small group builds commitment. However, there was a counter-tendency of participants wanting to leave projects to authorities as participation fatigue takes a toll on them. From a planning point of view, there is a need to manage the expectations of the participants and build social support among this group of participants.

Some success can be deduced from the intervention at Block 839. The converted void deck at Block 839—the Learning Link—with its community café (Palmwalk Café) and library attracted some 55 residents each day. On average, these residents stayed for about 20 min at the café, mingling and chatting with their neighbours. The converted space also sees more ground-up initiatives organised by the residents, such as pot luck sessions and cooking classes.

After the opening of the Palmwalk Café, a community garden was built in front of it near Block 839 (Fig. 6). In contrast to the participatory project that was initiated by the government, this garden was proposed and constructed by a resident, Mr. Chew Wei, who plays the role of the garden's guardian. He made clear his wish for it to resemble a park, and to be kept unfenced in order to enable easy access for the community. Despite several roadblocks to implement the plan, Mr. Chew persevered, and after a year of liaison with the Housing and Development Board, the National Parks Board, and the town council, the community garden was given the green light in October 2016 (Chua 2017). The garden became a beloved communal space among the residents for gardening activities and informal gathering, as well as for the passersby to stop and look on at the garden, all of which facilitated casual conversations. The produce grown from the garden is often shared among the residents at the Palmwalk Café as ingredients for food and drinks (Chua 2017). It is noteworthy to point out that Mr. Chew was one of the most active participants in the Tampines participatory project throughout the entire process in 2014-2015. Besides tending the garden, he frequently participates in various activities organised at the Palmwalk Café, and is an active volunteer in managing the café together with other residents.

The success of the Palmwalk Café and the community garden initiated by a local resident suggests an important shift from just participation to self-initiated action

Fig. 6 Community garden next to the Palmwalk Café. *Source* Im Sik Cho

as the café and the garden were in full operation at the time of study. These residents were actively involved in co-managing the community spaces as they began to contribute food and beverages and volunteer in other ways. This is a significant development as it illustrates the compassion and sharing principles that are fundamental in resilient neighbourhoods (Dubas 2011). The Palmwalk Café example also shows the importance of the organisational element (Residents' Committees) and governance structure in providing a contact point for the building of social capital, an essential element in resilient neighbourhoods (Luederitz et al. 2013; Allmark et al. 2014). The community garden which is now flourishing in synergy with the café is an example that shows a successful communal space that was co-created by the community through a community-initiated process and self-organised collective action.

The evidence from Figs. 3 and 4, and Table 1 suggests that while participation among residents is difficult and that there is fatigue involved in the participatory process, the Palmwalk Café case shows how, in Singapore's case, an established network developed in Residents' Committees (RCs) may provide the necessary structure and incentives for further individual initiatives like the one developed by Mr. Chew. Such organised networks ensure that individual members do not have to carry the burden of organising and can concentrate on specific activities. However, even these networks are unable to include hard-to-reach categories of residents, such as (i) parents with young children, (ii) schooling youths, and (iii) young adults who are building their careers and (need to) spend long hours at their workplace.

6 Conclusion: Towards Long-Term Social Sustainability and Resilience of Neighbourhoods

Resilient neighbourhoods are important social units in building the social fabric of the city. The propensity of local residents to participate is the first step in grassroots' efforts at city building and resilience. It is imperative not only to encourage participation but also to make sure that participation can be sustained. The level of community participation in the local Singaporean context has been largely related to informing, consulting, and involving residents, led by the authorities. Such exercises that have limited scope of participation often result in the local communities' perception that their impact on decision-making is limited. Furthermore, our interviews indicated that a large percentage of the community was of the opinion that the planning, improvement, and management of their housing estate would be adequately taken care of by the authorities. However, as Hollnsteiner (1976) argues, this may result in apathy and lack of interest and initiative among citizens to take on larger roles in neighbourhood design and planning, which go beyond daily matters.

The research project has attempted to foster active participation in planning neighbourhood spaces in order to create a sense of ownership and build strong(er) bonds among community members. However, community participation is complex, not static, and is heavily influenced by cultural, social, political, economic, and physical factors. Within Singapore's dense and multi-ethnic urban environment, residents tend to value their primary relations with their family and friends (from school and work), with residual attention being paid to neighbours. Moreover, the neighbourhood is seen as a place of rest and relaxation from the demands of school and work and, therefore, residents tend to guard their privacy. Residents understand and respect this need for privacy and keep interactions to a friendly casual greeting when encountering their neighbours. This superficial relationship between neighbours is also because neighbourhoods in public housing estates are well served by the town council, which provides efficient essential services and thus negating the need for residents to organise collective action to petition for better services. Consequently, there will be difficulties in sustaining collaborative involvement among neighbours in such contexts. In order to sustain a community project, the local government needs to create ways to hand it over to the community. A collaborative platform is crucial to (i) maintaining participation, (ii) growing and transforming neighbourhood resources, and (iii) mediating between various stakeholders and communities. A truly co-operative approach to neighbourhood planning should harness public assets by establishing adequate organisational structures where residents, professionals, and community groups work together, brainstorm, and create plans for improving their neighbourhoods in various creative ways (as demonstrated by Mr. Chew, for example). This will facilitate informed decision-making and move engagement beyond consultation towards more robust forms of cooperation and collaboration. This would cultivate a well-informed community that is better able to understand the decision-making process of planning, in addition to the consequences of these choices for Singapore society.

It is crucial to start with grasping the local circumstances in order to guarantee that the project responds directly to the community's actual needs, as well as to capitalise on existing neighbourhood resources and opportunities. Responding to the reality on the ground means (i) comprehending the specific demographic, cultural, socio-economic and political influences in the locality; (ii) establishing a common interest based on this understanding; and (iii) involving diverse views in the collaborative process. To facilitate collaboration with community stakeholders, a combination of various techniques ensures greater involvement from residents in terms of number and diversity. The research project has tested novel approaches to involving the community, with participation at the core of neighbourhood design for shared communal areas. However, achieving inclusivity remains a challenge. More emphasis should be placed on engaging everyone in a locality, including hard-to-reach groups or those with traditionally low involvement profiles. Fainstein (2013) argued that participation potentially makes policy-making more responsive to citizens' needs, but it also 'has weaknesses that can undermine its inclusivity and effectiveness'. Civic engagement raises the question 'of which citizens to involve' (Fainstein 2013).

Singapore's recent move towards a more inclusive style of leadership has encouraged greater local awareness about the importance of community involvement. This research project was launched in line with this evolving approach to neighbourhood planning; it has attempted to develop a participatory mechanism that fosters deeper social cohesion and community bonding by engaging the community and stakeholders in the planning process. It has also aimed to provide potential avenues for strengthening a community's capacity to initiate change and resolve shared problems through joint efforts, which is essential to supporting the city's long-term social sustainability and resilience.

The project has generated positive outcomes such as more ground-up initiatives organised by residents, increased use of newly created communal spaces and deeper neighbourly interactions taking place at these spaces. Sense of community is not just a social construct. It has a spatial dimension, as it is an amalgam of shared emotional contact through interaction with others in a particular place. The 'sense of community' encompasses a 'sense of place' attachment and a 'sense of membership' whereby a community member feels and shares a "right to belong" (Talen 1999). Although challenges remain which need to be addressed in order to build and sustain resilient neighbourhoods, this study is an attempt to initiate and explore offers of insights into possible mechanisms of participatory design, whereby potentials exist for various stakeholders and members of the community to be empowered and contribute to the planning process of their living environment, with the aim to enhance social interaction, sense of community, and resilience of the neighbourhood.

Moving beyond the immediate tangible outcomes, it is important to once again remind ourselves that community-building is a 'process' that needs to involve all stakeholders, rather than an end goal in itself (Talen 2000). Therefore, it is essential to create more opportunities for self-organisation, social learning and experimentation to empower the community, which many studies have pointed out as a key-strategy for building resilience (Goldstein 2009; Desouza and Flanery 2013 cited in Schauppenlehner-Kloyber and Penker 2016). 'The focus on strengthening problem-solving capacity endorses the view of community as a process', while 'community as artifact…is something that simply exists, something to be discovered' (Talen 2000). This view is relevant because it shifts the focus from any immediate result or ideal 'image' of community to the process of building social capital, which is indispensable for long-term social sustainability and resilience.

Acknowledgements This chapter is based on "Study on the Application of Design Recommendations to foster Community Bonding" (R-294-000-050-490) and "Study on the Design Recommendations to foster Community Bonding" (R-111-000-141-490), which were collaborative research projects between the National University of Singapore (NUS) and Singapore's Housing and Development Board (HDB) during 2014–2015. As principal investigators of the respective projects, the authors would like to thank the HDB for their support and funding.

References

Allen J, Cars G (2001) Multiculturalism and governing neighbourhoods. Urban Stud 38(12):2195–2209

Allmark P, Bhanbhro S, Chrisp T (2014) An argument against the focus on community resilience in public health. BMC Public Health 14(62)

Baker M, Coaffee J, Sherriff G (2006) Achieving successful participation, spatial plans in practice: supporting the reform of local planning. University of Manchester, London
Buchholz RA, Rosenthal SB (2004) Stakeholder theory and public policy: how governments matter. J Bus Ethics 51(2):143–153
Centre for Sustainable Asian Cities (CSAC) (2014) Study on impact of the built environment on community bonding (R-294-000-035-490), CSAC/HDB, Final report. NUS, Singapore
Centre for Sustainable Asian Cities (CSAC) (2015) Study on the application of design recommendations to foster community bonding (R-294-000-050-490), CSAC/HDB, Final report. NUS, Singapore
Cho IS (2016) HDB-NUS study on the social aspects of the built environment. Innov Mag—Mag Res Technol 15(1):47–54
Cho IS (2017) Changing approaches to community participation for social sustainability: neighbourhood planning in Singapore and Seoul. In: Caprotti F, Yu L (eds) Sustainable cities in Asia. Routledge, London, pp 226–235
Cho IS, Križnik B (2017) Community-based urban development. Evolving urban paradigms in Singapore and Seoul. Springer, Singapore
Cho IS, Nasution I, Lee J, Mascarenhas N (2017) Mechanisms to facilitate community participation in Singapore's neighborhood planning framework. J Archit Plann Res 34(4):320–335
Chua XT (2017) Friendships flourish at Tampines community garden, café. TODAY
Dubas R (2011) Lessons in building resilient neighbourhoods: reflections on the PROUT Institute Community SEED Program. Communities 153:17–19
Fainstein SS (2013) The limits to public engagement. Urban Solutions 3:80–83. Centre for Liveable Cities, Singapore
Hee L, Khoo LM (2013) Engaging communities: lessons from around the world. Urban Solutions 3:57–65
Hee L, Ooi GL (2003) The politics of public space planning in Singapore. Plann Perspect 18(1):79–103
Holden M (2011) Public participation and local sustainability: questioning a common agenda in urban governance. Int J Urban Reg Res 35(2):312–329
Hollnsteiner MR (1976) People power: community participation in the planning and implementation of human settlements. Philippine Stud 24(1):5–36
Innes JE, Booher DE (2004) Reframing public participation: strategies for the 21st century. Plann Theory Pract 5(4):419–436
Irvin RA, Stansbury J (2004) Citizen participation in decision making: is it worth the effort? Public Adm Rev 64(1):55–64
Jha AK, Miner TW, Stanton-Geddes Z (2013) Building urban resilience: principles, tools, and practice (directions in development). World Bank, Washington DC
Luederitz C, Lang DJ, Von Wehrden H (2013) A systematic review of guiding principles for sustainable urban neighborhood development. Landscape Urban Plann 118:40–52
MacDonald C (2012) Understanding participatory action research: a qualitative research methodology option. Can J Action Res 13(2):34–50
Ooi GL (2004) Future of space: planning, space and the city. Eastern University Press, Singapore
Putnam R (2000) Bowling alone: the collapse and revival of American community. Simon and Schuster, New York
Sanoff H (2000) Community participation methods in design and planning. Wiley, Chichester
Schauppenlehner-Kloyber E, Penker M (2016) Between participation and collective action—from occasional liaisons towards long-term co-management for urban resilience. Sustainability 8:664
Selenger D (1997) Participatory action research and social change. Cornell University, New York
Talen E (1999) Sense of community and neighbourhood form: an assessment of the social doctrine of new urbanism. Urban Stud 36(8):1361–1379
Talen E (2000) The problem with community in planning. J Plann Lit 15(2):171–183
Taylor RB (1996) Neighbourhood responses to disorder and local attachments: the systemic model of attachment, social disorganization, and neighbourhood use value. Sociol Forum 11(1):41–74

Bringing Arts Closer to Local Communities: Spatial Opportunities and Impacts on Community Bonding

Zdravko Trivic

Abstract 'Arts and Culture Nodes' is an initiative launched by Singapore's National Arts Council (NAC) in 2012 to bring quality arts closer to where people live, foster regular arts engagement, enrich public life and build stronger community bonds. The key mechanism employed was not only creating a comprehensive network of partnerships with artists and various neighbourhood institutions, such as community and recreational clubs, libraries and non-profit organisations across the island, but also activating a range of neighbourhood public spaces as more informal arts and culture venues. This chapter first outlines the research conducted in five Singaporean neighbourhoods to investigate spatial, social and participation impacts of the 'Nodes' initiative. Guided by the 'Neighbourhood Arts and Culture Impact Assessment' (NACIA) framework and the conceptual lenses of placemaking and cultural ecology, this chapter then discusses the roles and capacities of arts and culture initiatives in building sustainable community arts development and resilient local urban communities.

Keywords Community arts · Arts impacts assessment · Creative placemaking · Neighbourhood cultural ecology · Community bonding

1 Introduction

The last few decades witnessed a revived interest around the world in studying the role of community arts and culture in shaping the perception and use of urban environment and creating stronger neighbourhood identities and local communities. This restored interest brought the shift from predominant focus on economic impacts of arts and culture towards a range of 'non-economic' outcomes (e.g., Dwyer et al. 2000; Wood 2009). Many studies concluded that engagement with arts could generate numerous positive socio-cultural outcomes, including enhancing personal development

Z. Trivic (✉)
Department of Architecture, National University of Singapore, Singapore, Singapore
e-mail: akizt@nus.edu.sg

C.-H. Leong and L.-C. Malone-Lee (eds.), *Building Resilient Neighbourhoods in Singapore*, Advances in 21st Century Human Settlements,
https://doi.org/10.1007/978-981-13-7048-9_7

(e.g., Matarasso 1997), health and wellbeing benefits (e.g., Stuckey and Nobel 2010), improved educational performance (e.g., Hetland and Winner 2001), and increased social capital and community empowerment (e.g., Stern and Seifert 2009), among others.

Government authorities and funding agencies worldwide have understood the extent of such impacts and embraced arts initiatives as means to achieve greater societal goals. However, arts impact investigations invite some criticism as they press arts and culture organisations to justify their existence based on outcomes that have very little to do with their key mission statements, i.e. producing high-quality arts and cultural experiences. Yet, the empirical research assessing 'non-economic' impacts of arts and culture is still relatively scarce, with little agreement regarding the terminology and evaluation mechanisms employed. The reasons may lie in the nature of such impacts, as they are often subjective, difficult to measure, take time to accrue and resist statistical causality (e.g., Belfiore and Bennett 2010).

This chapter outlines a study conducted in five housing neighbourhoods in Singapore to (i) investigate their capacities to build stronger arts and culture ecology and (ii) capture and evaluate impacts of arts events, under 'Arts and Culture Nodes' strategy. Moreover, the chapter aims to discuss the key findings with respect to the role of community arts in building resilient local urban communities. Section 2 briefly introduces the key milestones in Singapore's cultural development, focusing on the growth of community-centred arts and culture initiatives over the last decade. In Sect. 3, the approach and methods employed in this research are outlined, including a conceptual background, a study of neighbourhood spatial opportunities and a study of impacts of arts and culture, based on proposed 'Neighbourhood Arts and Culture Impact Assessment' (NACIA) framework. In Sect. 4, the key spatial, social and participation impacts observed by the study are presented and discussed in reference to building community resilience. Finally, the concluding Sect. 5 reflects on the overall impacts of the 'Nodes' strategy, its capacity to harness available spatial and human resources for and through the arts and to empower local communities through regular arts participation.

2 Arts and Culture Development in Singapore

Over the last ten years, arts and culture development in Singapore has experienced a substantial shift, which embraced community arts to promote local identity, racial harmony and social cohesion (Lee and Sim 2016; Lim 2014; Low 2015), in addition to the predominant focus on premium venues and events typically associated with city centre (Fig. 1).

Early cultural development in Singapore reflected the need to build distinctive national identity, especially after gaining independence in 1965, and the key infrastructure that was needed to support its aspiration of becoming a 'global city' in the 1980s and the 1990s (Centre for Liveable Cities [CLC] 2017). The key milestones were the 'Report of the Advisory Council on Culture and the Arts' released in 1989,

Fig. 1 Timeline of arts and culture development in Singapore. *Source* By author; based on "Timeline: Planning and development for culture and the arts" (CLC 2017, pp. 106–115)

which formulated recommendations towards building a culturally vibrant society by 1999, and the birth of the National Arts Council (NAC) in 1991.

The next milestone was the 'Renaissance City Plan I', released in 2000 by the Ministry of Information and the Arts (MITA), which articulated key strategies for turning Singapore into a 'global cultural hub', and marked a substantial shift from cultural infrastructure to strengthening the sense of belonging, cohesion and resilience among Singaporeans (CLC 2017; MITA 2000). Two more Renaissance City Plans have been released afterwards. The 'Renaissance City Plan II' (Ministry of Information, Communications and the Arts [MICA] 2005) articulated an industry-focused approach to cultural development focusing on (i) building up new skills and capabilities within the arts and cultural sector, and (ii) encouraging arts/culture-business partnerships. Boosting an integrated ecosystem for the arts, culture and heritage sectors and building community engagement were the main strategies established by the subsequent 'Renaissance City Plan III' (MICA 2008). Community engagement strategy involved integration of arts and culture and neighbourhood facilities, such as schools, libraries, community centres, malls and parks. In the long run, MICA envisioned partnering with the Housing Development Board (HDB) and the Urban Redevelopment Authority (URA) to explore new models of town planning that would integrate arts and culture into the physical and experiential environment of local neighbourhoods.

Finally, in 2012, the Ministry of Culture, Community and Youth (MCCY) further reinforced the community-focused arts and culture policies by launching the 'Arts and Culture Strategic Review' (ACSR) that charted the course for Singapore's cultural development until 2025 (MCCY 2012). Primarily driven by the private and people sectors, and supported rather than led by the MICA, this review reflected the changing aspirations and demands of the arts community to more actively involve the public in the processes of learning, practice, arts creation and co-creation, performance and organisation, rather than solely arts consumption. In result, the ACSR engendered 'Community Engagement' and 'Arts and Culture Education' master plans, which aimed to empower the public in shaping Singapore's cultural development, boost ground-up initiatives and safeguard its long-term sustainability.

2.1 Community Arts and Culture Development in Singapore

Various community arts and culture initiatives have been introduced across Singapore since 2012. One such initiative is the 'PAssionArts Movement' launched by the People's Association (PA) to energise local neighbourhoods and enable collaborations between professional artists, community talents and grassroots organisations through arts (MCCY 2016). HDB's 'Community Week' (started in 2012) and URA's 'Our Favourite Place' (piloted in 2013) are some of the other initiatives that supported community-initiated arts and creative placemaking projects.

In 2008, NAC started the 'Arts for All' initiative comprising several schemes to support community arts and collaborations between artists, amateur arts groups, corporations, community partners and voluntary welfare organisations (MCCY 2016).

'Arts in Your Neighbourhood' is one such effort, involving a series of arts and culture programmes at various locations across the island, typically throughout March and November of every year. Among other initiatives are: 'Silver Arts' targeted at senior citizens, 'NAC-ExxonMobil Concert in the Park' that brings concerts to Singapore Botanical Gardens, 'Busking Scheme' offering individuals and groups to exhibit their talents on streets of Singapore, and 'Got To Move' that brings free dance performances and classes island-wide (NAC 2018).

2.1.1 'Arts and Culture Nodes' Strategy

In 2012, NAC initiated the 'Arts and Culture Nodes' strategy (or 'Nodes' strategy), which is the focus of this study. The main aim of this strategy was to establish a comprehensive island-wide network of arts touch-points in the local neighbourhoods. The objectives were not only to provide all residents greater access to quality arts, but also to bring buzz to neighbourhood public spaces, enrich public life, build stronger local identity, and increase the opportunities for community participation and bonding (NAC 2014). Accordingly, the 'Nodes' strategy expands beyond the traditional arts venues, such as museums, arts galleries or theatres. Instead, NAC partners with a range of neighbourhood institutions, such as community clubs, libraries, recreational clubs and non-profit organisations, and embraces various neighbourhood public spaces as more informal arts venues. By 2017, the 'Nodes' strategy established 15 arts and culture nodes island-wide, with a vision of creating 20 nodes by 2020 (Fig. 2). Since 2012, they have engaged more than 200 artists and arts groups, offered over 1000 arts programmes, and reached out to over 240,000 attendees (NAC 2017).

According to the 'National Population Survey on the Arts' (NAC 2015), the overall interest in the arts among Singaporeans has significantly increased since 2005. In 2015, both arts attendance and participation doubled in comparison to 2013, which might be the outcome of NAC's strategic efforts, but also of an unusually high number of events offered to celebrate Singapore's 50th birthday that year. Residents described attending cultural events as enjoyable experiences and good opportunities for spending quality time with others. These were the main reported reasons for arts participation.

3 Approach and Methods

With specific reference to 'Arts and Culture Nodes' strategy, this study investigates the capacities of five Singaporean neighbourhoods (as highlighted in red in Fig. 2) to create culturally rich and vibrant environments, and generate positive impacts on local communities through diverse arts experiences. Approach and methods combine ethnographic research, spatial mapping, surveys, interviews and focus group discussions to gather both quantitative and qualitative insights about the quality and suitability of arts venues, and assess the impacts of arts and culture initiatives.

Fig. 2 Arts and culture nodes in Singapore (2017). *Note* Names highlighted in red represent the five nodes investigated in this study. *Source* By author

The study was conducted in two phases: (i) literature review and neighbourhood spatial opportunity analysis, and (ii) study of spatial, social and participation impacts of arts events on local neighbourhoods and communities.

3.1 Literature Review and Conceptual Background

The conceptual framework is built upon a comprehensive review of theories, policies and initiatives pertinent to the quality of public space and impacts of arts and culture, with specific focus on concepts of 'placemaking' and 'cultural ecology'.

3.1.1 'Placemaking' and 'Creative Placemaking'

'Placemaking' is a process of deliberate shaping of an environment to support social interaction, improve quality of life of local communities, empower residents through public engagement, and boost civic pride (Silberg et al. 2013). While extensively used since 1990s to describe community-centric approach to public space design, planning and management, placemaking emerged in the 1960s from the ground-breaking ideas against systematic destruction of people-friendly spaces. Its pioneers were William H. Whyte and Jane Jacobs, who emphasised the shift from physical aspects of spaces towards social life.

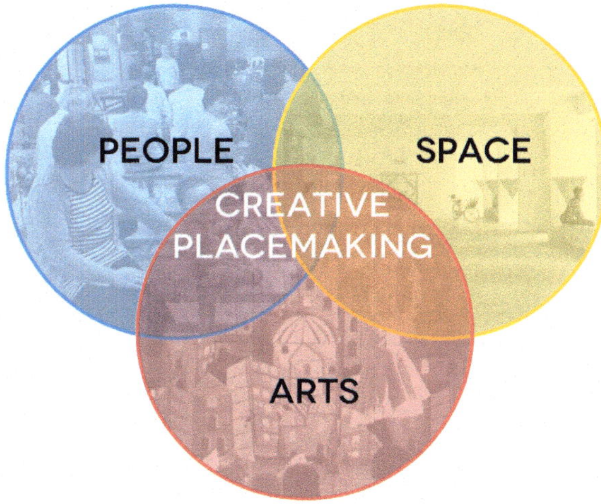

Fig. 3 Concept of creative placemaking: Symbiotic arts-space-people relationship. *Source* By author

While placemaking may involve an array of arts and non-arts related actions aimed at revealing and enhancing the unique characters of physical and social forms of local communities (Artscape 2016a; Project for Public Space 2013), 'creative placemaking' is a process of bringing positive impacts to the community through deliberate and strategic creation of symbiotic relationships between arts and culture, space and people (Fig. 3) (e.g., Markusen and Gadwa 2010; Pollock and Paddison 2014). While creative placemaking does not necessarily imply permanent changes to spaces, it always involves enhancing the 'sense of belonging and attachment' to both neighbourhood and community. Creative placemaking can thus be a useful tool for augmenting the existing place identities or stimulating the creation of new ones while animating local communities and their specific values through arts and culture at neighbourhood places that can serve as key collective anchor points.

3.1.2 Arts and Culture Ecology

The design quality of public space directly affects the nature and quality of optional and social activities taking place there, like arts and culture activities (Gehl 1987). The main premise is that the success of community arts programmes considerably depends on careful selection of spaces and partners in the neighbourhood and on understanding how local residents use them.

Neighbourhood arts and culture ecology refers to a comprehensive network of 'hard' (space) and 'soft' (people) infrastructure established to support and sustain regular arts activities (Fig. 4). 'Hard' infrastructure comprises arts-related facilities, such as spaces for arts consumption and promotion (e.g., performance, exhibition), education and training (e.g., rehearsal, practice), and production (e.g., arts studios, workshops). It also includes various supporting spaces available in the neighbourhood

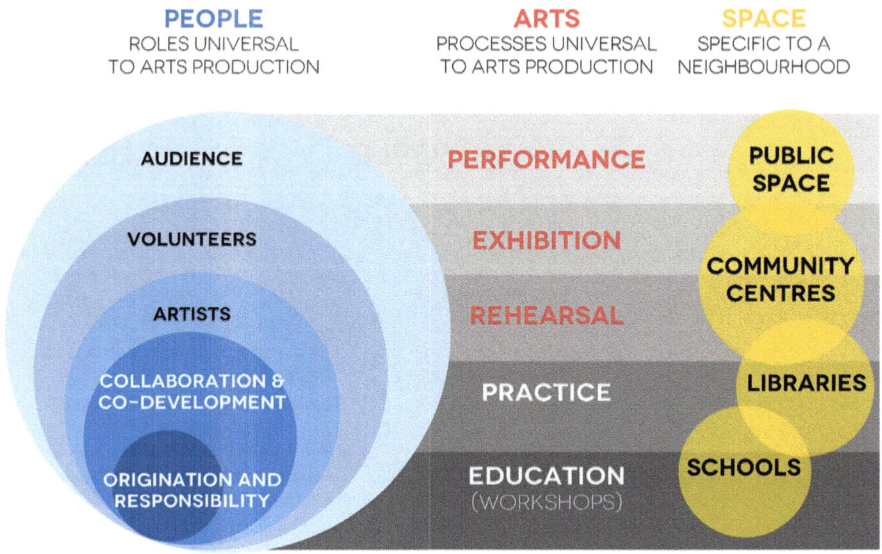

Fig. 4 Elements of neighbourhood arts and culture ecology. *Source* By author

that can provide critical social and economic support to the arts (e.g., commercial and work spaces, cafés, schools, public spaces) (Artscape 2016b; GovHK 2015). In such a way, different arts-related activities could take place at a variety of spaces that may not be designed or dedicated specifically for the arts, and also attract and engage diverse stakeholders in an informal manner. 'Soft' infrastructure, refers to a blend of formal and informal networks between individuals, community institutions and non-profit organisations (Stern 2014; Taylor 2008). Employing the potentials of symbiotic clustering of all creative neighbourhood resources is crucial for facilitating arts collaboration and cross-fertilisation of creative plans and ideas, as it offers a wider range of concentrated and complementary intersections for face-to-face interaction, networking, and cultural and social exchange among a varied mix of stakeholders (Artscape 2016b).

3.2 Study of Neighbourhood Spatial Opportunities

A systematic study of spatial opportunities and overall design qualities of available spaces was conducted to guide the choice of the most suitable nodes in local neighbourhoods for synergetic community arts and culture activities. Five neighbourhoods, covering 400 m radius area around each arts and culture node, were investigated. Each node differs in respect to its spatial capacities, engagement of public spaces, goals and target audiences, and brings a distinct flavour to its surrounding neighbourhood.

Kallang Community Club (CC), one of the oldest nodes, primarily caters to families with children and senior citizens. It organises regular monthly events with local professional artists and arts hobby groups, both within its own premises (open courtyard, indoor multi-purpose hall and practice rooms) and in neighbourhood public spaces (e.g., open plaza at Boon Keng train station and Bendemeer mall) (Fig. 5). *Woodlands Regional Library (WRL)* and *SAFRA Jurong* (recreation club) focus on their regular patrons, young adults and families with children, and offer a variety of performing arts events and workshops, primarily within their own indoor premises (e.g., auditoriums, exhibition spaces, atria and function rooms) (Fig. 6).

Tanglin Halt is a sheltered public space located next to the train station, supermarket and a food court, and it is activated by a non-profit organisation 'My Community' (Fig. 7). 'My Community' focuses mainly on local artists and intangible cultural heritage of the Queenstown neighbourhood, offering a range of programmes, from music and theatre performances to hands-on workshops in outdoor venues. *Bedok Town Square* is a large sheltered space next to public transportation hub, shopping mall and a hawker centre and hosts large-scale events operated directly by the NAC (Fig. 7).

Fig. 5 Arts venues at Kallang: CC Courtyard (left), CC Auditorium (middle) and Boon Keng Plaza (right). *Source* By author

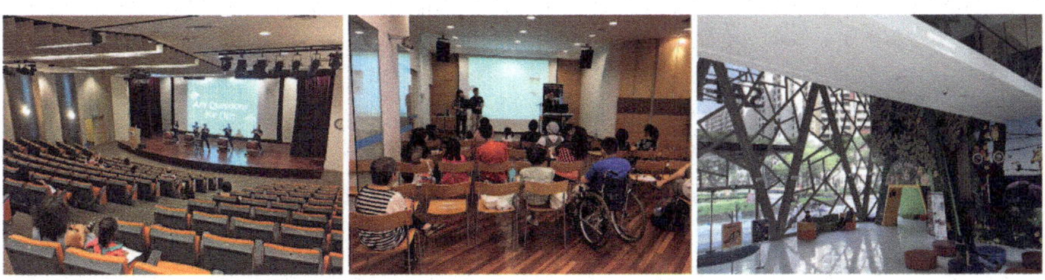

Fig. 6 Indoor venues at WRL: Auditorium (left) and Programme Zone (middle) and SAFRA Jurong: Atrium (right). *Source* By author

Fig. 7 Outdoor venues used by 'My Community': Tanglin Halt (left), Community Plaza (middle), and Bedok Town Square (right). *Source* By author

3.2.1 Spatial Opportunity Analysis

Spatial opportunity analysis consists of two steps: (1) on-site ethnographic investigation and spatial mapping of all publicly accessible spaces, and (2) map-overlay analysis and evaluation.

In Step 1, two layers of data are documented: general spatial conditions (scale, enclosure and type of ground surface) (Fig. 8), land use and intensity of pedestrian movement and stationary activities, employing snapshot counting and tracing techniques (Gehl and Svarre 2013) (Fig. 9). Generated maps are then overlapped to reveal and shortlist spaces that are the most suitable for holding arts events.

In Step 2, shortlisted spaces are evaluated against the criteria for good public space and suitability for arts venues, based on Urban Space Framework by Cho et al. (2016), other relevant literature (e.g., Carmona et al. 2010; GovHK 2015; Shaftoe 2008) and interviews with event organisers and artists. Criteria included physical design attributes—'hardware' (e.g., accessibility, flexibility and supporting infrastructure) and socio-perceptual qualities—'software' (e.g., diversity of uses, sociability and place character), with some indirect considerations of operational and organisational conditions—'orgware' (e.g., time and noise regulations, logistics and participation). This detailed evaluation assists event organisers to understand the strengths and weaknesses of available spaces and synergise them with specific requirements of their arts programmes.

As a result of superimposing the maps from Step 1, four main categories of neighbourhood spaces emerged (Fig. 10). Category A involves spaces close to both transient and stationary activities, such as those next to public transportation hubs or shopping malls. They are generally suitable for broadening and diversifying the arts audience. Category B includes spaces that are away from major pedestrian paths, but are well-utilised, such as basketball courts or green lawns. These more intimate spaces may be suitable for small-scale events and activities to deepen residents' engagement with the arts. Type C comprises spaces found near pedestrian movement paths but have no substantial stationary activity, such as pavilions or covered walkways. In these spaces, arts could encourage residents to pause and engage with each other,

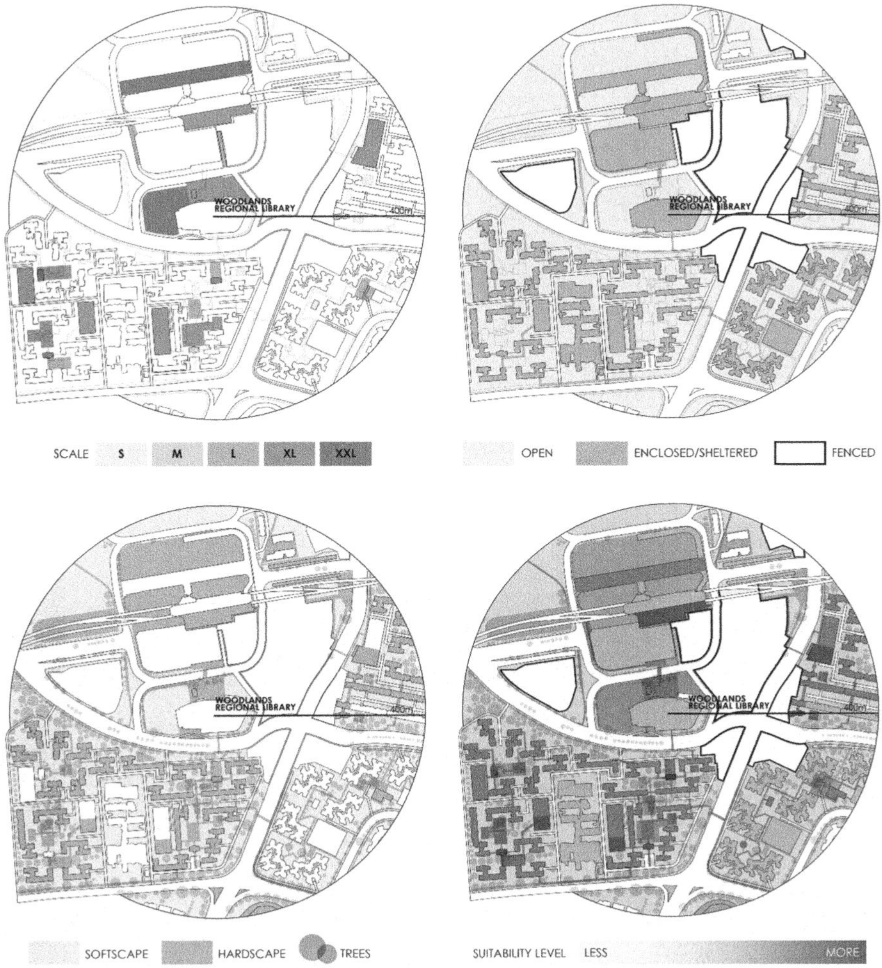

Fig. 8 Step 1: Mapping general spatial conditions in Woodlands neighbourhood: scale (top left); enclosure (top right); ground surfaces (bottom left); overlay of three layers—darker shades indicate spaces of greater suitability (bottom right). *Source* By author

or advertise events held in other less accessible venues. Finally, category D involves spaces that are away from both pedestrian movements and stationary activity, such as amphitheatres or roofs of multi-storey car parks. Arts can play an important role in activating such under-utilised neighbourhood spaces.

Fig. 9 Step 1: Mapping land use (left) and pedestrian movement and stationary activities (right) in Woodlands neighbourhood. *Source* By author

Fig. 10 Space categories in Woodlands neighbourhood (from left to right): Category A—Civic plaza next to train station and shopping mall; Category B—Well-utilised basketball court; Category C—Under-utilised open space alongside covered walkway; Category D—Accessible multi-storey carpark roof-top. *Source* By author

3.3 Study of Impacts of Arts and Culture

Impacts refer to a range of outcomes that may result from arts and culture activities, whether short-, intermediate- or long-term, 'intrinsic' or 'instrumental' (e.g., Carnwath and Brown 2014; Holden 2004; Keeney and Korza 2015). Intrinsic impacts are formed and experienced through subjective encounters with arts. They may not be immediately visible nor easy to measure yet they can make long-lasting influence on individuals. Instrumental impacts, on the other hand, refer to secondary effects (e.g., economic, social or health outcomes), which could also be triggered by non-arts means.

3.3.1 'Neighbourhood Arts and Culture Impact Assessment' (NACIA) Framework

Impact study consisted of pilot and final stages. In the pilot stage, an initial 'Neighbourhood Arts and Culture Impact Assessment' (NACIA) framework was proposed. Works by McCarthy et al. (2004), Dunphy (2015) and Brown and Novak-Leonard (2013), among others, provided valuable insights into the key domains, indicators and methods for measuring and evaluating arts impacts in an interconnected and holistic manner. A preliminary database of nearly 200 impact indicators was narrowed down based on insights gathered from the focus group discussions (FGDs) with the node partners and the residents.

The final NACIA framework comprises two key domains (attributes and impacts), 8 dimensions and 20 indicators (Fig. 11). Attributes describe and quantify the conditions related to arts events and should be distinguished from the impacts, although they may affect certain outcomes and indicate some aspects of success of nodes' efforts.[1]

With the main focus on capturing and evaluating spatial (hardware), social (software) and participation attributes and impacts, NACIA framework was applied at 19

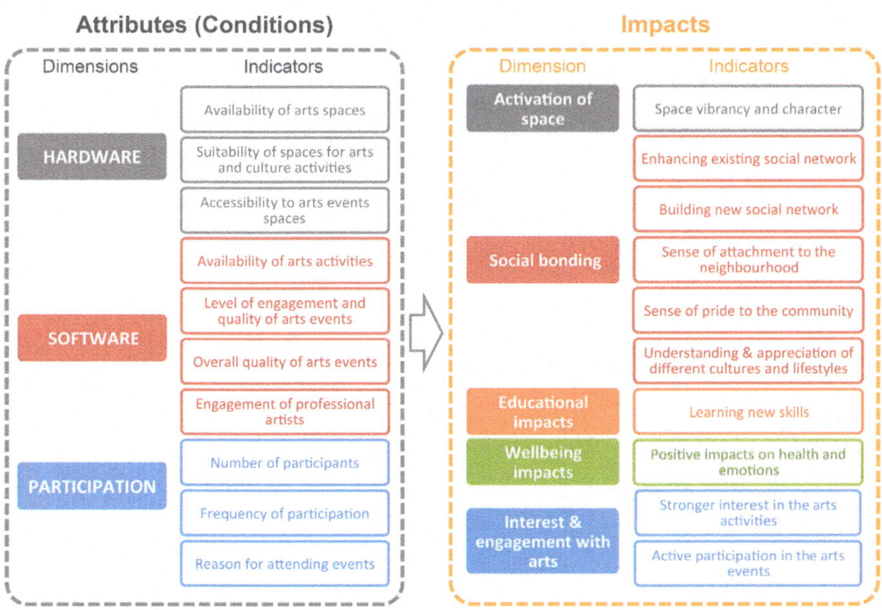

Fig. 11 NACIA Framework. *Source* By author

[1]For instance, the number of participants may depend on type of an arts event and the spatial capacity of an event venue (small-scale vs. large-scale, intimate vs. mass event). While it might somewhat indicate the event's success, it may not be a fair indicator of an impact, although it may influence the perception of place vibrancy, for example.

arts events in five neighbourhoods, including 242 pilot surveys and 300 final surveys, 10 post-event discussions with a total of 58 residents and 8 interviews with event organisers and artists.

3.3.2 Analysis

Survey analysis. Data obtained through surveys were analysied using the Statistical Package for the Social Science (SPSS) software, based on nominal and/or ordinal variables (see e.g., Greasley 2008). Over 60% of the survey samples consisted of 30–59-year-old female adults of Chinese ethnicity. An even number and proportion of age, gender and ethnic groups across the five neighbourhoods could not be achieved, as some art events clearly catered to specific or small audiences.

Focus group discussion (FGD) analysis. The analysis of FGD transcripts adopts five stages of qualitative analysis proposed by Ritchie and Spencer (1994): (a) familiarisation with data; (b) thematic framework identification; (c) indexing; (d) charting; and (e) mapping and interpretation. QDA Miner qualitative data analysis software is used to distil the issues residents mentioned the most frequently. The main themes identified include arts participation and arts venues (Fig. 12), and impacts of the arts and culture events (Fig. 13).

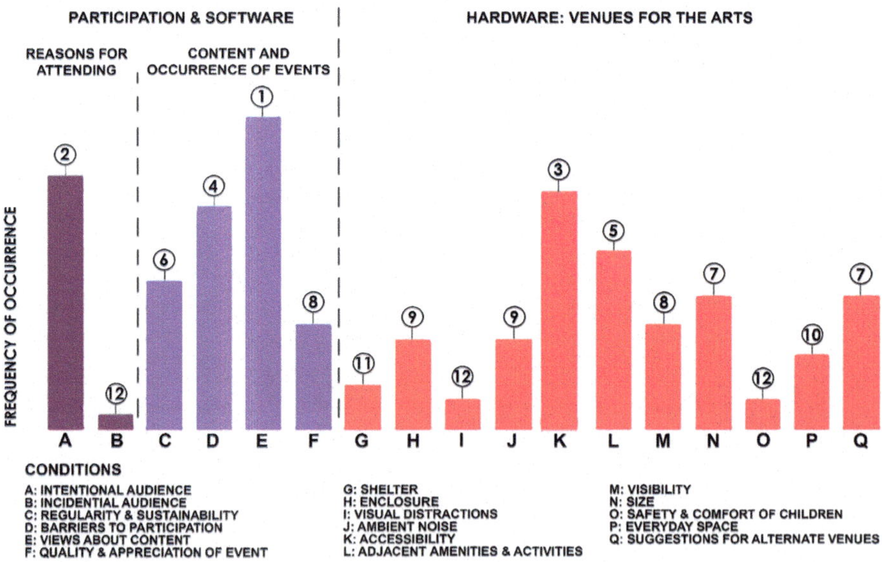

Fig. 12 Key FGD themes—Frequency of participation and venue. *Source* By author

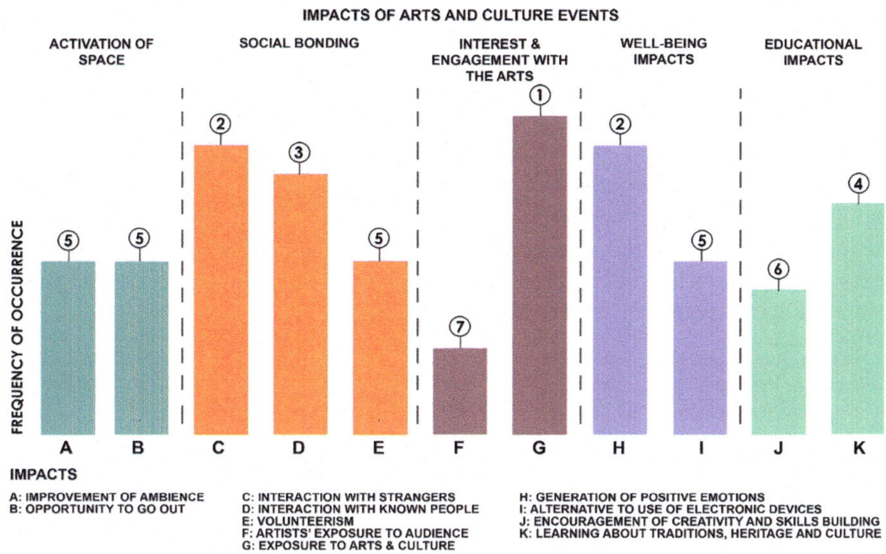

Fig. 13 Key FGD themes—Frequency of impacts on arts and culture. *Source* By author

4 Key Findings and Discussion

The following key findings are outlined according to NACIA framework and discussed through the conceptual lenses of creative placemaking and cultural ecology.

4.1 Attributes—What Can Space Do for the Arts?

4.1.1 Space (Hardware)

Space availability and utilisation. The node partners employ a range of spaces for arts and culture activities, yet with different degrees and approaches to space utilisation and activation. Such spaces range from formal indoor venues (e.g., auditoriums or multi-functional halls) to more informal settings (e.g., open plazas). Utilisation of spaces within and beyond nodes' premises indicate their capacities and undergoing processes of establishing a successful neighbourhood arts and culture ecology. For example, while WRL and SAFRA Jurong tend to keep arts activities within their own premises, Kallang CC and My Community@Queenstown seem more active in outreach by hosting their programmes at various locations across their neighbourhoods. However, the capacities to engage spaces in the neighbourhood more intensively are not exhausted, including, for instance, schools' and other institutions' indoor and outdoor premises that can provide conducive environments for cultural events. Pilot survey revealed that the residents would like to see arts events in various neighbourhood spaces, from civic centres, schools, foodcourts and shopping malls to parks, basketball courts and void-decks. Good utilisation of available spatial resources may be critical for the success and sustainability of arts and culture in the neighbourhood.

Suitability for arts and culture events. The majority of surveyed participants found all investigated event venues "suitable" or "very suitable" for arts activities. Interestingly, outdoor venues were found somewhat more suitable than indoor venues. According to FGD findings, accessibility and adjacent amenities are the most important venue features that affect people's decision to attend an event, followed by size of space, visibility, presence of shelter[2] and ambient noise (Fig. 12).

> **It's sheltered**, (...) the community centre hall, so **you don't have to worry about the weather and also there's air-conditioning,** so whether rain or shine, (...) [it] is conducive for an art performance. (Kallang CC FGD, 21 Jan' 2017)

Residents' perception of space suitability varies depending on type of event. While the proximity to high pedestrian traffic flows generally suits performances, such spaces are also noisy and distracting, and therefore less conducive for activities that require more focused attention (e.g., workshops).

> I **don't think this is a very conducive place**. We are being **surrounded by four things: the road, the supermarket, the food-court and the coffee-shop**. (...) I find that it is problematic as it is **too noisy**. (...) with four walls of **distraction**, personally, I find that it is a bit tough. (Tanglin Halt FGD, 24 Jun' 2017)

The majority of participants are in favour of having arts events closer to their homes, especially the elderly and mothers with young children, for whom travelling a long distance to attend an arts event can be challenging. Moreover, the presence of adjacent amenities, such as grocery stores, food courts or MRT stations, makes a venue more attractive and convenient, as it allows the residents to combine their arts attendance with other activities.

> **You can watch a performance and buy groceries here**; you can just put the things down, **it's convenient**. (Tanglin Halt FGD, 25 Feb' 2017)

In some instances, adjacent amenities, such as the food court at Tanglin Halt, not only attract higher number of participants but can also act as an extension of audience seating area (Fig. 14).

4.1.2 Programme (Software)

Diversity of programmes. Overall, there is a good diversity of arts programmes offered across all nodes, ranging from more intimate workshop events to music, dance

---

[2]Due to Singapore's tropical climate characterised by high air temperature, sun exposure, humidity and rainfall levels (especially during the rainy season in December and January), the protection from such harsh weather conditions is an important factor that affects the use of open public spaces and, therefore, their suitability for as arts and culture venues.

Fig. 14 Food court adjacent to Tanglin Halt plaza. *Source* By author

and theatre performances. Diversity of programmes attracts different user groups and provides opportunities for social interaction through arts. However, the regularity and frequency of such programmes vary across the five nodes (Table 1), depending on their individual capacities (e.g., availability of space and human resources).

Table 1 Spatial capacity, diversity of programmes and frequency of events across nodes

Nodes	WRL	SAFRA Jurong	Kallang CC	Bedok Town Square	Tanglin Halt
Spatial capacity	Indoor premises (auditorium and activity rooms)	Indoor premises (atrium, activity rooms and restaurant)	Indoor and outdoor premises (multi-purpose hall, courtyard, practice rooms and public spaces)	Single outdoor space (open public plaza)	Outdoor spaces (open public plazas)
Programmes	Performing arts, workshops, visual arts	Performing arts and workshops	Performing arts and arts classes (interest groups)	Performing arts	Performing arts and workshops
Scale of events	Large and small scale	Medium and small scale	Large, medium and small scale	Large scale	Medium and small scale
Frequency of events	Monthly (irregular)	Quarterly (irregular)	Monthly (regular)	Quarterly (irregular)	Monthly (irregular)

Source By author (information collated from NAC and node partners)

Quality of arts events. Regardless of the type of events and the engagement of professional artists, the majority of participants (90% on average) find community arts and culture events of overall very good quality, enjoyable and engaging. Some FGD participants, however, questioned whether some of the arts events could be considered "high quality arts" or simply entertainment.

> **I don't know whether you consider that art. They have parties** (…). So they bring in some singing, musicians and things like that. (SAFRA Jurong FGD, 17 Dec' 2016)

Audience's perception of the quality of arts events is influenced by a number of factors, such as the entrance fee, suitability of the venue, atmosphere, arts content, relationship to performers, etc. Findings from both surveys and FGDs clearly show that intentional attendees, the so-called 'art lovers', have higher expectations than incidental audience. Free events are also associated with the lower quality by some residents.

4.1.3 Participation (Orgware)

Attendance. Overall, the 'Nodes' strategy proved successful in attracting new audience and enriching opportunities for residents across Singapore to engage with arts, as nearly 70% of surveyed participants were the "first-timers". For a sustained cultural development, however, repeated audience is equally important. Greater percentage of repeated audience was found in nodes that hold arts events more frequently and engage public spaces, such as Kallang CC (45.2%). Event attendance varies across nodes mainly due to location and type of event, time and duration of the event, means of advertising and ticketing. For instance, while performance events generally attract larger crowds, workshop events often allow only limited number of pre-registered participants.

Active participation. Active participation in the arts is one of the key indicators of community building and creative placemaking. However, the majority of participants (about 80%) prefer 'passive' engagement. Only 24.2% and 8.3% of the participants expressed their desire to be more actively engaged in future events as volunteers or performers, respectively. The most common barriers reported for active participation are time constraints and lack of incentives.

> **For volunteering, I wouldn't be that keen as I'd prefer watching** (…), because at this age we are more practical (…). **To sacrifice your weekends or work during the night, you'd expect higher pay instead of free labour**. (Bedok FGD, 02 Jun' 2017)

The repeated audience and 'art lovers' are more likely to actively participate in the arts than the first-time and incidental participants. Active participation provides the means for people to express their relationship with their physical and social environment and can be fruitful vehicle for community empowerment and boosting self-initiated arts programmes, which are critical for sustaining rich arts ecology in the neighbourhood.

4.2 *Impacts of Arts and Culture*

Overall findings indicate that the 'Nodes' strategy has generated various positive impacts on local neighbourhoods and communities, but the levels of impacts vary across the nodes.

4.2.1 Hardware Impacts—What Can Arts Do for Space?

Activation of space and vibrancy. More than 90% of participants acknowledged that arts events brought vibrancy and sense of identity to their neighbourhood spaces. Arts events make the residents see familiar spaces from a new perspective and provide opportunities to explore unfamiliar neighbourhoods.

> It [arts event] **creates good atmosphere and promotes a light-hearted, joyous mood.** (Tanglin Halt FGD, 25 Feb' 2017)

For many participants, vibrancy and atmosphere come from the crowd and short-term changes occurring in space, such as temporary seats, banners, stage, shelter, lighting or music, which make the space feel inviting and festive. It was also observed that the residents like to use such temporary amenities both before and after the events in an informal manner for further socialisation or resting. Such a capacity of arts events to enrich the public life beyond the actual duration of the performance should be further explored.

> **It was actually the music that attracted me to this place**. I was actually going home (…) and then I realised it was something live so I thought it was quite cool (…). [It is] quite interesting and **makes the whole place more lively**. (Kallang FGD, 21 Jan' 2017)

> Very good, **I like it, there's better atmosphere**. (…) If only there's performance every week. (…) This takes place only once every 2 to 3 months. (Tanglin Halt FGD, 26 Feb' 2017)

Attachment to neighbourhood. 81.4% of participants reported stronger attachment to their neighbourhood after attending an arts event. Statistical analysis also found correlations between the sense of attachment and the quality of events, engagement and enjoyment levels and the perception of space suitability.

4.2.2 Social Impacts (Software)—What Can Arts Do for Community?

Community bonding. Arts and culture events proved to be important settings for both strengthening audience's existing social bonds and networks and building new ones. Social interaction is the prerequisite for developing stronger social cohesion and social capital. The majority of residents attended the events with family members or friends to spend some quality time together (Table 2).

On the other hand, only negligible number of residents joined the event together with their neighbours, and one third of participants did not communicate with

Table 2 People whom respondents attend events with

Nodes	WRL		SAFRA Jurong		Kallang CC		Bedok Town Square		Tanglin Halt		Total	
Myself/Alone	27	33.3%	1	4.4%	48	40.0%	52	39.4%	51	30.4%	179	34.2%
Family members	37	45.7%	20	87.0%	46	38.3%	49	37.1%	79	47.0%	231	44.1%
Friends	21	25.9%	3	13.0%	28	23.3%	32	24.2%	34	20.2%	118	22.5%
Neighbours	0	0.0%	0	0.0%	2	1.7%	2	1.5%	4	2.4%	8	1.5%
Children (below 12)	9	11.1%	4	17.4%	7	5.8%	6	4.6%	21	12.5%	47	9.0%
Others	3	3.7%	0	0.00%	0	0.0%	1	0.8%	2	1.2%	6	1.2%
Total answers	97	100.0%	28	100.0%	131	100.0%	142	100.0%	191	100.0%	589	100.0%
Valid sample	81	119.8%	23	121.7%	120	109.2%	132	107.6%	168	113.7%	524	112.4%

Source By author

anyone during the event. However, 44.08% of respondents stated that they spoke to people they already knew or even made new friends (22.52%), which is encouraging (Table 3).

> Yeah, I think any kind of event creates interest, interest **creates topics that you can make small talk about**. If you're in a lift with a fellow you can say "Oh hey, I saw you at so and so event. (Kallang FGD, 21 Jan' 2017)

> **The person sitting behind me today was from my floor**, he was there for the performance as well, **and thus we could talk, which we normally would not do**. (Tanglin Halt FGD, 26 Feb' 2017)

The level of social interaction during an event depends on a number of factors, including the nature of the venue (formal or informal) and the level of interactivity of the performance. Informal venues and workshop activities seem to prompt spontaneous interaction better than formal venues.

> **In a formal space** (…) **you don't take note of who sits near you** (…) but if it's informal you're like 'Who's as busybody as me?' (Kallang FGD, 21 Jan' 2017)

Therefore, participation in the arts events provide a range of opportunities for the residents to progress along all stages of 'neighbouring', from passive face-to-face contact and mutual recognition to shared experience, further interaction and, finally, friendship (Grannis 2009). In such a way, arts and culture can actively contribute to community bonding and building of social capital, which are the crucial elements of social sustainability and resilient communities.

Pride to community and appreciation of different cultures. Nearly 90% of attendees felt higher sense of pride towards their community after attending arts events. Self-achievement, supporting family members or friends performing at an event and celebrating particular cultural heritage are the most obvious "pride triggers". Nearly 90% of surveyed participants "agreed" or "strongly agreed" that participating in arts and culture programmes improved their understanding and appreciation of different cultures and lifestyles. Social cohesion, mutual appreciation and support, beyond mere tolerance and coexistence, are particularly important for multi-cultural societies like Singapore.

4.2.3 Educational Impacts

As perceived by the residents, arts participation can boost creativity, learning new skills and acquiring new knowledge in a playful manner.

> It's unexpected (…). **I'm experiencing things I might not know, or listen to things that I never heard before, which is something new**. (Kallang FGD, 21 Jan' 2017)

> So how can we shape the mind-set that **arts do help in building up your soft skills**, which will help to contribute to whatever career your child chooses ultimately? (SAFRA Jurong FGD, 17 Dec' 2016)

> **Learning history through play.** Give them the history, such as Queenstown, butterfly flats, through play - building blocks. It is not just creativity as in what you want to build. (…) that is learning through play. (Tanglin Halt FGD, 24 Jun' 2017)

Table 3 Level of communication with other people at event

Nodes	WRL		SAFRA Jurong		Kallang CC		Bedok Town Square		Tanglin Halt		Total	
No, I didn't speak to anyone	35	33.3%	9	4.4%	51	40.0%	64	39.4%	67	30.4%	226	34.2%
Yes, I spoke to people I already know	27	45.7%	11	87.0%	45	38.3%	42	37.1%	59	47.0%	184	44.1%
Yes, I made new friends	16	25.9%	2	13.0%	27	23.3%	22	24.2%	43	20.2%	110	22.5%
Total answers	78	100.0%	22	100.0%	123	100.0%	128	100.0%	164	100.0%	520	100.0%
Valid sample	76	119.8%	22	121.7%	120	109.2%	127	107.6%	169	113.7%	509	112.4%

Source By author

4.2.4 Wellbeing Impacts

Health and wellbeing are at the heart of sustainable and resilient communities. Attending arts events affect the participants' mood and overall wellbeing positively. This is one of the most immediate impacts, and almost 90% of surveyed participants agreed that they felt happier and healthier after attending an art event. Some of the positive emotions triggered, as mentioned by the participants, include: a sense of relaxation, joy, inspiration and appreciation of life. Different age groups may experience wellbeing impacts in different ways. For the elderly, attending an event is an opportunity to go outdoors, be active and make friends. For parents, attending arts events is an activity the whole family can engage in together and therefore a preferable alternative to being absorbed with electronic devices at home.

> Any form of event will (…) generate a certain interest and I think this interest is very important because **it's true when you say we live a very mundane life. A lot of people say 'Singaporeans got no life, we are not happy', so this brings the life back to us**. (Kallang FGD, 21 Jan' 2017)

> **Singing is an activity which keeps one healthy** (…). When I visit the doctor, he will ask me, 'Did you exercise?'. I reply, 'No, but I sing. I sing while walking.' (…) I am 76 years old. (…) The [health] report stated that I can stay healthy by going for such activities. (…) I came from Woodlands. I took the MRT here alone. (Kallang FGD, 28 Apr' 2017)

Manifestation of such positive emotions are also seen in some participants who expressed their enjoyment in a spontaneous manner, such as dancing together with the performers.

4.2.5 Participation Impacts (Orgware)—What Can Community Do for the Arts?

More than 70% of participants expressed the desire to pursue additional arts activities, such as creative hobbies or arts classes, upon participating in an arts and culture event. Some residents also expressed their aspirations to be mentored by the community artists and have their work showcased together at a community venue. Only through sustained mentorship by the artists and arts event organisers, the individual members of the community can build the necessary skills for initiating community arts events on their own. Workshop participants tend to develop stronger interest in arts after attending an art event in comparison to performance participants and they are also more likely to be actively engaged in the future events.

5 Conclusion

This study first explored the spatial capacities of the local housing neighbourhoods to develop a stronger neighbourhood arts and culture ecology and proposed steps for the systematic selection of the most suitable public spaces for community arts events and activities. By employing a comprehensive 'Neighbourhood Arts and Culture Impact Assessment' (NACIA) framework, the study then evaluated spatial, social and participation impacts of community arts events on local urban communities and revealed the capacities of arts and culture initiatives in boosting sustainable community arts development and resilient local urban communities.

Overall, it can be concluded that the 'Nodes' strategy is a systematic and comprehensive approach to community arts development that complements other NAC and non-NAC initiatives to promote arts in the heartland of Singapore. The strategy has proved effective in overcoming the physical barriers of bringing arts closer to where people live and play. The node programmes are overall well received by the audience. Each node may have different target groups and different outreach strategies, but as a whole they have succeeded in attracting diverse participants in terms of age, ethnicity and work profile.

The findings showed that the 'Nodes' strategy generated a variety of positive impacts on local neighbourhood and communities, including (i) bringing vibrancy to public spaces, (ii) boosting positive emotions and community bonding, (iii) building stronger appreciation for the arts and different cultures, (iv) learning new skills, and (v) being active. It can be noticed that the nodes operating at the neighbourhood level, engage wider ranges of neighbourhood spaces (arts ecology), offer diverse programmes on more frequent and regular basis, and seem to generate overall higher impacts. Such impacts go beyond the rather superficial quantity of arts programmes or "feel good" indicators, and although predominantly short-term, tend to bring important collective meanings and values to the community.

Community participation and volunteerism are essential for building creative and resilient urban communities. The resources needed to organise frequent arts events at the neighbourhood level present a challenge that can be met with a stronger neighbourhood arts ecology. Moreover, the involvement of professional artists can play a critical role in the transference of skills and mentorship to amateur neighbourhood performers to boost ground-up self-initiated activities and volunteerism. This is particularly apparent at the Kallang CC node, with the most comprehensive and mature arts and culture ecology among the five neighbourhoods (Fig. 15). Kallang CC not only employs a range of outdoor and indoor spaces within and beyond its premises for arts performances, rehearsals and practice, but has also established the interdependence between professional art programmes, amateur community arts practice and voluntarism.

Crafting arts and culture events that reflect on material and symbolic values of specific neighbourhoods and communities also shows to be an important element. Employing a number of public spaces across Queenstown neighbourhood, 'My Community' engages both local residents and the wider public with the neighbourhood-specific heritage, which results not only in high enjoyment but also in willingness to volunteer and pursue other creative activities as well as in strengthening the existing social bonds and building new ones.

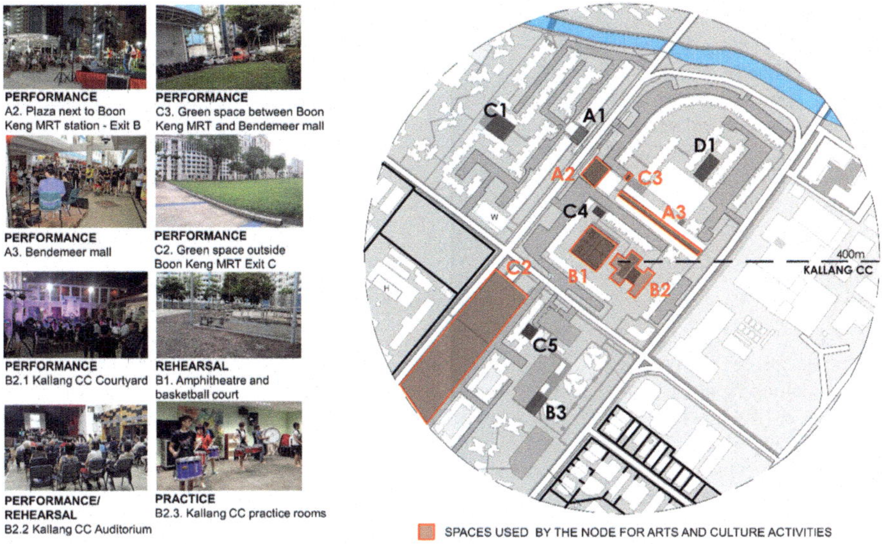

Fig. 15 Arts and culture ecology in Kallang neighbourhood. *Source* By author

Cultural sustainability has only recently gained full attention and formal recognition in the frameworks of sustainable development and resilience, as the "forth pillar", alongside economic, social and environmental dimensions (e.g. James et al. 2015; Kagan 2007; Soini and Birkeland 2014; United Cities and Local Governments 2010; United Nations 2017).[3] In many aspects, this study confirmed the capacities of the 'Arts and Culture Nodes' strategy to build sustainable arts and culture development within the local neighbourhoods and develop stronger and resilient local communities. Strategic use of spatial and human resources, promoting civic engagement, triggering sense of belonging and ownership, fostering social cohesion, inter-generational and inter-cultural bonding reflect some of such capacities. However, in order to boost full potentials of community arts and culture to make greater societal impacts, arts itself still needs to find its resilience within the community. Some of the most essential tasks, initiated by the 'Nodes' strategy, are: (i) to understand and synergise all material and human assets available in the community, (ii) to provide adequate infrastructure and platforms for empowering the community to take an active creative initiative (community engagement, ground-up initiatives and volunteerism), and (iii) to develop collective community vision and partner network.

[3]The concept of sustainable development, developed in the late 1980s, initially proposed three key pillars of sustainability, namely: economic growth, social inclusion and environmental balance (United Nations General Assembly 1987). Although advocated since the early 2000s, United Cities and Local Governments in 2010 for the first time proposed culture as the fourth pillar of sustainable development, which is now widely accepted.

Acknowledgements The author would like to express his appreciation to Beng Kiang Tan, Nina Mascarenhas, Quyen Duong, and Rita Padawangi for their invaluable contribution to this project. Research project "Arts and Culture Nodes in the Heartlands of Singapore: Exploring Strategies, Spatial Opportunities and Impacts (Phase 2)" (R-294-000-061-490) discussed in this chapter, was conducted at the Centre for Sustainable Asian Cities (CSAC), School of Design and Environment (SDE), National University of Singapore (NUS), in collaboration with and funded by the National Arts Council (NAC), Singapore.

References

Artscape (2016a) Approaches to creative placemaking. http://www.artscapediy.org/Creative-Placemaking/Approaches-to-Creative-Placemaking.aspx. Accessed 20 Apr 2018

Artscape (2016b) Place matters. D.I.Y. Creative placemaking. http://www.artscapediy.org/Creative-Placemaking/Place-Matters.aspx. Accessed 20 Apr 2018

Belfiore E, Bennett O (2010) Beyond the "toolkit approach": arts impact evaluation research and the realities of cultural policy-making. J Cult Res 14(2):121–142. https://doi.org/10.1080/14797580903481280

Brown AS, Novak-Leonard JL (2013) Measuring the intrinsic impacts of arts atten-dance. Cult Trends 22(3–4):223–233. https://doi.org/10.1080/09548963.2013.817654

Carmona M, Health T, Oc T, Tiesdell S (2010) Public places urban spaces: the dimensions of urban design. Elsevier, Amsterdam; Oxford

Carnwath JD, Brown AS (2014) Understanding the value and impacts of cultural experiences: a literature review. Arts Council England. https://www.artscouncil.org.uk/sites/default/files/download-file/Understanding_the_Value_and_Impacts_of_Cultural_Experiences.pdf. Accessed 20 Apr 2018

Centre for Liveable Cities (2017) Urban systems studies: a city of culture: planning for the arts. Centre for Liveable Cities, Singapore

Cho IS, Heng CK, Trivic Z (2016) Re-framing urban space: urban design for emerging hybrid and high-density conditions. Routledge, New York

Dunphy K (2015) A holistic framework for evaluation of arts engagement. In: Macdowall L, Badham M, Blomkamp EM, Dunphy K (eds) Making culture count: the politics of cultural measurement. Palgrave Macmillan, London, pp 243–263

Dwyer L, Mellor R, Mistilis N, Mules T (2000) A framework for assessing 'tangible' and 'intangible' impacts of events and conventions. Event Manag 6:175–189. https://doi.org/10.0000/096020197390257

Gehl J (1987) Life between buildings: using public space. Van Nostrand Reinhold, New York

Gehl J, Svarre B (2013) How to study public life. Island Press, Washington; Covelo; London

GovHK (2015) Hong Kong planning standards and guidelines: Chapter 3 community facilities. Planning Department, The Government of the Hong Kong Special Administrative Region. http://www.pland.gov.hk/pland_en/tech_doc/hkpsg/full/ch3/ch3_text.htm#9. Accessed 20 Apr 2018

Grannis R (2009) From the ground up: translating geography into community through neighbour. Princeton University Press, Princeton

Greasley P (2008) Quantitative data analysis using SPSS: an introduction for health and social science. McGraw-Hill/Open University Press, Maidenhead; New York

Hetland L, Winner E (2001) The arts and academic achievement: what the evidence shows. Arts Educ Policy Rev 102(5):3–6. https://doi.org/10.1080/10632910109600008

Holden J (2004) Capturing cultural value: how culture has become a tool of government policy. Demos, London

James P, with Magee L, Scerri A, Steger MB (2015) Urban sustainability in theory and practice: circles of sustainability. Routledge, London

Kagan S (2007) Art and sustainability: connecting patterns for a culture of complexity. Transcript Verlag, Bielefeld

Keeney KP, Korza P (2015) Assessing arts-based social change endeavors: controversies and complexities. In: Stephenson MO, Tate AS (eds) Arts and community change: exploring cultural development policies, practices and dilemmas. Routledge, New York and London, pp 186–211

Lee J, Sim JL (2016) Arts-based community engagement in Singapore: success stories, challenges, and the way forward. In: Hersey LN, Bobick B (eds) Handbook of research on the facilitation of civic engagement through community art. IGI Global, pp 391–411

Lim B (2014) Rethinking community arts in Singapore. J Theatrex Asia 1(1):15–22

Low F (2015) Community cultural development—why construct an Asia perspective? In: Low F (ed) Art with communities: practices and possibilities. Partridge Publishing, Singapore

Markusen A, Gadwa A (2010) Creative placemaking. Markusen Economic Research Services and Metris Arts Consulting, Washington. https://www.arts.gov/sites/default/files/CreativePlacemaking-Paper.pdf. Accessed 20 Apr 2018

Matarasso F (1997) Use or ornament: the social impact of participation in the arts. Comedia, Stroud. http://www.culturenet.cz/res/data/004/000571.pdf. Accessed 20 Apr 2018

McCarthy KF, Ondaatje EH, Zakaras L, Brooks A (2004) Gifts of the muse reframing the debate about the benefits of the arts. RAND Corporation. https://www.rand.org/content/dam/rand/pubs/monographs/2005/RAND_MG218.pdf. Accessed 20 Apr 2018

Ministry of Culture, Community and Youth (MCCY) (2012) The report of the arts and culture strategic review. MCCY, Singapore. https://www.nac.gov.sg/dam/jcr:1b1765f3-ff95-48f0-bbf9-f98288eb7082. Accessed 20 Apr 2018

MCCY (2016) An active citizenry through sports and broadening access through arts and heritage. MCCY, Singapore. https://www.mccy.gov.sg/news/speeches/2016/Apr/active-citizenry-through-sports-and-broadening-access-through-arts-heritage.aspx. Accessed 20 Apr 2018

Ministry of Information and the Arts (MITA) (2000) Renaissance city report: culture and the arts in renaissance Singapore. MITA, Singapore. https://www.nac.gov.sg/dam/jcr:defaf681-9bbb-424d-8c77-879093140750. Accessed 20 Apr 2018

Ministry of Information, Communications and the Arts (MICA) (2005) Renaissance city plan II. MICA, Singapore. https://www.nac.gov.sg/dam/jcr:c6cbd326-a895-4d28-b91b-a0b8ae715878. Accessed 20 Apr 2018

MICA (2008) Renaissance city plan III. MICA, Singapore. https://www.nac.gov.sg/dam/jcr:18cf2883-7907-4938-9931-384333e210ce. Accessed 20 Apr 2018

National Arts Council (NAC) (2014) More community arts and culture nodes bring the arts to the heartlands. NAC, Singapore. https://www.nac.gov.sg/media-resources/press-releases/More-Community-Arts-and-Culture-Nodes.html. Accessed 20 Apr 2018

NAC (2015) Population survey on the arts. NAC, Singapore. https://www.nac.gov.sg/dam/jcr:9a606902-09d8-4505-b880-92e5da253de6. Accessed 20 Apr 2018

NAC (2017) Enjoy year-round arts programmes at NAC's Community Arts and Culture Nodes. NAC, Singapore. https://www.nac.gov.sg/media-resources/press-releases/NAC-Community-Arts-and-Culture-Nodes.html. Accessed 20 Apr 2018

NAC (2018) Engaging the public. NAC, Singapore. https://www.nac.gov.sg/whatwedo/engagement/artsforall.html. Accessed 18 June 2018

Pollock VL, Paddison R (2014) On place-making, participation and public art: the Gorbals, Glasgow. J Urban: Int Res Placemaking Urban Sustain 7(1):85–105. https://doi.org/10.1080/17549175.2013.875057

Project for Public Spaces (2013) All placemaking is creative: how a shared focus on place builds vibrant destinations. https://www.pps.org/article/placemaking-as-community-creativity-how-a-shared-focus-on-place-builds-vibrant-destinations. Accessed 20 Apr 2018

Ritchie J, Spencer L (1994) Analysing qualitative data. In: Bryman A, Burgess RG (eds) Qualitative data analysis for applied policy research. Routledge, London, pp 173–194

Shaftoe H (2008) Convivial urban spaces: creating effective public places. Earthscan, Sterling; London

Silberg S, Lorah K, Disbrow R, Muessig A (2013) Places in the making: how placemaking builds places and communities. MIT Department of Urban Studies and Planning, Massachusetts. https://dusp.mit.edu/sites/dusp.mit.edu/files/attachments/project/mit-dusp-places-in-the-making.pdf. Accessed 20 Apr 2018

Soini K, Birkeland I (2014) Exploring the scientific discourse of cultural sustainability. Geoforum 51:213–223. https://doi.org/10.1016/j.geoforum.2013.12.001

Stern MJ (2014) Measuring the outcomes of creative placemaking. In: The role of artists and the arts in creative placemaking, 30–31 May 2014, Baltimore, MD—Symposium Report. Goethe-Institut and EUNIC, Washington DC, pp 84–97. http://www.goethe.de/ins/us/was/pro/creative_placemaking/2014_Symposium_Report.pdf. Accessed 20 Apr 2018

Stern MJ, Seifert SC (2009) Civic engagement and the arts: issues of conceptualization and measurement. University of Pennsylvania, Philadelphia. https://animatingdemocracy.org/sites/default/files/CE_Arts_SternSeifert.pdf. Accessed 20 Apr 2018

Stuckey HL, Nobel J (2010) The connection between art, healing, and public health: a review of current literature. Am J Public Health 100(2):254–263. https://doi.org/10.2105/ajph.2008.156497

Taylor DG (2008) Magnetizing neighborhoods through amateur arts performance. Urban Inst. https://www.urban.org/sites/default/files/publication/32131/411784-Magnetizing-Neighborhoods-through-Amateur-Arts-Performance.PDF. Accessed 20 Apr 2018

United Cities and Local Governments (2010) Culture: fourth pillar of sustainable development. http://www.agenda21culture.net/sites/default/files/files/documents/en/zz_culture4pillarsd_eng.pdf. Accessed 22 Nov 2018

United Nations (2017) New urban agenda. http://habitat3.org/wp-content/uploads/NUA-English.pdf. Accessed 22 Nov 2018

United Nations General Assembly (1987) Report of the world commission on environment and development: our common future. United Nations. http://www.un-documents.net/wced-ocf.htm. Accessed 22 Nov 2018

Wood EH (2009) An impact evaluation framework: local government community festivals. Event Manag 12:171–185. https://doi.org/10.3727/152599509789659768

Place Familiarity and Community Ageing-with-Place in Urban Neighbourhoods

John Chye Fung

Abstract Successful ageing requires the active involvement of society and the concomitant self-responsibility for keeping healthy in old age. Resilience, both at the individual and community level, is an important component of successful ageing. A life-course perspective of active ageing which considers the temporal evolvement of an elderly person over a lifetime, and not simply a snapshot account of old age, is needed to integrate time and place as a singular space relative to older adults' experience of their living environment. With this perspective, beyond the usual agency framing of individual resilience at old age, community support of active ageing becomes quintessential as a part of the collective resilience of urban neighbourhoods. Here, place familiarity is crucial in engendering psychosocial and urban resilience, which affects seniors' wellbeing. At the individual level, this community-oriented approach to active ageing operates as a wellspring to draw upon against adversities, and collectively as part and parcel of resilience building at the neighbourhood level. This paper illustrates the close association between place and active ageing in contributing to older adults' experience of highrise, high-density urban neighbourhoods that are common to Singapore. It advances the concept of collective urban and psychosocial resilience.

Keywords Active ageing · Resilience · Place-making · Neighbourhood · Field theory

1 Introduction

Contemporary discourse on population ageing invariably advances the notion that ageing need not, indeed must not, be passive. In social gerontology, *successful ageing* —a concept that is popular among policy makers—advocates active involvement

J. C. Fung (✉)
Centre for Ageing Research in the Environment, National University of Singapore, Singapore, Singapore
e-mail: akifjc@nus.edu.sg

C.-H. Leong and L.-C. Malone-Lee (eds.), *Building Resilient Neighbourhoods in Singapore*, Advances in 21st Century Human Settlements,
https://doi.org/10.1007/978-981-13-7048-9_8

in society and the concomitant self-responsibility for keeping healthy in old age. However, scholars who contest such a reductive attempt at a singular definition of the complexities of ageing argue that there is no 'gold standard' of how one should grow old. The related term *active ageing* alludes to self-initiated care and social participation in old age, promoting environments and activities that optimise the opportunities for elderly people to stay healthy by participating in social, cultural, spiritual, economic and civic life (World Health Organization [WHO] 2002). It is similarly challenged; albeit, to a lesser extent, as individuals' disposition towards activity differs. In contrast, the concept of *resilience* has emerged as an appropriate alternative to frame the discourse of successful ageing.

Ageing should adopt a life-course perspective by considering the temporal evolvement of an elderly person over a lifetime, and not simply promote a snapshot account of old age. The proposition of this essay is that an equivalent *lifespace*[1] understanding of ageing is needed to integrate time and place as a singular space relative to older adults' experience of their living environment. It argues for a commons approach to resilience based on field theory, beyond the usual agency framing of individual resilience at old age. Community support of active ageing is quintessential to collective resilience of urban neighbourhoods. Here, place familiarity is crucial in engendering psychosocial and urban resilience, which affects seniors' wellbeing.

Recent research in different knowledge areas has focused on the emerging notion of resilience as a disciplinary idea, resulting in many definitions and understanding although it is commonly agreed as a positive and desirable trait. While a multidisciplinary approach to urban resilience has been advocated to examine resiliency in psychological, social, ecological, engineering, infrastructural and economic terms (Chelleri and Olazabal 2012), there is little elucidation of what and how such intersection could be attained. Against this milieu, the two key interconnected strands that intersect with ageing and place are psychosocial and urban resilience. The former involves an individual's resistance of and coping with declining abilities in old age, while the latter is primarily framed as environmental sustainability. Missing here is the potential of place-making in supporting shared community values and collective actions to enhance the process of ageing in high-density environments. Psychosocial-urban resilience is most crucial and pertinent to rapid ageing populations at the meso-scale, high-density urban neighbourhood level where activities of daily living occur.

In the light of global trends such as population ageing and a shift from a pathogenic-pharmacologic view of health to an enriched model advancing a socio-ecological approach to wellbeing and quality of life, this essay discusses place familiarity and psychosocial-urban resilience in relation to the concept of *community ageing-with-place*. The transformation of a place is intricately connected with an evolving community and thus, active ageing occurs *with* rather than *in* place.

[1]In field theory, Kurt Lewin used *lifespace* to describe "the order of coexisting facts in a psychological or social situation" (Lewin and Lorsch 1939, p. 401). This essay expands its origin as a psychosocial attribute to include the geophysical environment and spatial experiences. It further advocates a collective multi-agent dimension beyond the individual agency view of *lifespace*.

By synthesising time and space as a single continuum, place familiarity acts as memory anchors for older adults in high-density urban neighbourhoods, thereby fostering resilience. Everyone in the community shares these experiences of place, albeit with difference nuances, hence effectively a collective resilience of the commons. Place familiarity is a factor contributing to both collective resilience and community ageing-with-place is examined under the ambit of neighbourhood planning.

This chapter advances a community-oriented approach to active ageing and resilience via the related concepts of *collective resilience* and *community ageing-with-place*. The author argues that *community ageing-with-place* is a more accurate term about growing together in a congruent manner to attain both individual and collective resilience. The notion of *resilience* is discussed in Sect. 2, supported by a theoretical framing using field theory in Sect. 3. For the elderly, resilience operates at both the individual level as a wellspring to draw upon against adversities, and collectively, as community support is needed for active ageing in an urban environment. Sections 4 and 5 examine the role of commons in community ageing-with-place; while Sects. 6 and 7 illustrate, using examples from Singapore, the key factor that affects ageing and resilience in a high-density context—*place-familiarity*. Section 8 lists the salient planning and design considerations of future urban neighbourhoods, a topic that concludes the chapter.

2 Resilience as a Critique of Successful Ageing

The landmark MacArthur Foundation Study of Successful Aging advanced *successful ageing* as encompassing three key components: (i) low risk of disease and disease-related disability, (ii) maintenance of high mental and physical function, and (iii) continued engagement with life (Rowe and Kahn 1998). Since then, the concept's positive slant to what was deemed a problematic phenomenon has captured the imaginations of social gerontologists and policy makers alike. Various models of successful ageing were postulated: Health Promotions Model (Rowe and Kahn 1997, 1998), Selective Optimization with Compensation (Baltes and Baltes 1990), Positive Aging Model (Valiant 2004) and Preventive and Corrective Proactivity Model (Kahana and Kahana 1996).

Conversely, critiques of successful ageing argue against the limitation of individual freedom and choice by emphasising health promotion behaviour (Riley 1998), the unquestioned use of the term without defining its inherent values (Holstein and Minkler 2003), the lack of a holistic concept in awareness of ageing and spiritual component (Moody 2005), and the lack of total life history perspective of positive ageing or gerotranscendence (Tornstam 2005).

Offering a contrasting paradigm that promotes resilience in old age as the preferred quest than successful ageing, the premise here is a more holistic and inclusive approach beyond health promotion imperatives. Cultivating resilience is deemed more accessible to older adults by disregarding socio-cultural conditions and physical or cognitive abilities, and thus is a better model than successful ageing

(Harris 2008). In developmental psychology (Yates and Masten 2004), resilience is a trait of children who develop well despite immersion in high-risk environments (Garmerzy 1973; Werner and Smith 1982). Resilience is an observable pattern of behaviour (Luthar and Zelazo 2003; Masten and Powell 2003), arises in children as an ordinary, normal function of human adaptive system (Masten 2001), and is affected largely by support from family and community (Luthar and Zelazo 2003). Resilience is assessed on two fundamental conditions: presence of an adversity to overcome, and the affected person performing well (Masten and Coatsworth 1998). It is a continuing human capacity to adapt and survive adversity, and could be commonly ascribed as "the ability to bounce back" (Harris 2008).

Among chronic illnesses associated with ageing, dementia is the most devastating to both the afflicted sufferer and his family. Therefore, resilience is a requisite trait for countering this debilitating condition. Most persons with dementia continue living purposefully after overcoming the initial distress of diagnosis (DeBaggio 2003; Harris 2002). If supportive psychosocial environment is present, people afflicted with Alzheimer's disease even at a late stage of the illness are able to maintain capabilities (Sabat 2001). Current dementia care philosophy reflects the importance of personhood as espoused by Kitwood (1997). This shift from pharmacological medicalisation to a more holistic psychosocial approach first articulated by Lyman (1989) and then advanced by Kitwood (1997) as person-centred care involves positive environment, human values, communication, culture and activities. Individual resilience includes the attributes of competence, adversity, assets, risk, protective factors and vulnerabilities (Harris 2008). The Nun Study (Snowdon 1997, 2003) lends hope to the possibility of averting the onset of dementia through environmental and behavioural adaptations, despite the presence of biomarkers that predisposes a person to the disease. It undergirds the current belief that dementia could be prevented, if not at least its onset impeded or delayed, through suitable lifestyle and environmental interventions.

Notwithstanding the contestation between resilience and successful ageing as the desired model for older adults, such studies invariably adopt a human agency perspective of individual wellbeing. Yet, resilience should be broadened as a systemic trait to include the collective commons by enlarging its definition and scope to include psychosocial resilience of a community, and urban resilience of a place. This refocus from singular human agency to diverse commons encapsulates a socio-ecological worldview, where a sustained system is integral and quintessential to individual survival. Resilience is meaningless if the psychosocial ecosystem fails systemically.

3 A Field Theory Perspective of Collective Resilience

Since time immemorial, animals have developed deep ecological ties with the environments they inhabit, primarily for survival and species propagation. Place familiarity is entrenched in establishing a home base and a means of staying alive. Even nomadic groups or tribes depend on place familiarity whilst traversing vast terrains stretching hundreds of square kilometres. Evolutionary processes through natural selection have embedded geospatial cognition and environmental adaptation in our survival skillsets. The relation between environment and gene expression is argued in the science of epigenetics. That the environment shapes our intrinsic ability to cope with life's stresses and to flourish entails resilience. The on-going Dunedin Study (Poulton et al. 2015) overturned earlier preconceptions that our propensity for exhibiting certain behaviours is determined solely by genetic makeup at birth—the 'nature' argument of human development. Instead, it showed that the environment plays an equal role in nurturing the expression or not, of one's genetic predisposition. Many fields now accept that the environment—physical, biochemical, social and virtual—facilitates or inhibits human behaviours and wellbeing. In the recent field of telomere science, Nobel laureate Elizabeth Blackburn established that the environment plays an important role in advancing the *healthspan* of people. Being safe and feeling safe in the neighbourhood are linked to a healthier lifespan. Physiologically, ageing is a reduction in cellular functional capacity to respond to injuries and environmental stimuli (Blackburn and Epel 2017). Individual resilience could be enhanced if the environment is supportive and engenders positive emotions, allowing a person to flourish.

Medical and social sciences of ageing often cite the environment as an important factor influencing physiological, cognitive and behavioural outcomes. Although research in social gerontology deepened our understanding of the psychosocial aspects of ageing, the related field of environmental gerontology is less well developed despite decades of interest (Diaz Moore 2014). Oftentimes, the environment was conceived as a contextual scaffold to support other arguments. For instance, while there is agreement that familiarity of place contributes to creation of supportive environments for long-term care, there is little evidence on how this is achieved through physical planning and design. Architectural and urban theories tend to advance place-making at a conceptual level, such as phenomenology, spatial experience, sensory perception and cognition, rather than empirically.

Many attempts were made at defining what constitutes *the environment* especially in relation to health; such as the therapeutic environment model comprising natural, built, social and symbolic aspects (Gesler 1992); salutogenesis model of health supportive environment that promotes a sense of coherence through comprehensibility, manageability and meaningfulness (Antonovsky 1979); total environment encompassing physical, natural, social, behavioural dimensions (Smith et al. 1999); categorisation in gerontology as physical, personal, small-group, suprapersonal and social (Lawton 1999); and as a constantly contested social construct of community (Delanty 2010).

In environmental gerontology, psychologist Powell Lawton postulates the Ecological Theory of Ageing (Lawton and Nahemow 1973) emphasising human adaptation to changes in the environment. Salient to this theory is an allusion to the environment acting as a press that imposes on people, whose ability or competence to adapt and overcome such forces varying with individuals. Beyond the comfort and challenge zones, the environment is either over- or under-whelming, leading to maladaptive behaviours. Lawton extended Kurt Lewin's field theory to include person-environment transaction as a factor affecting human behaviour. Lewin's field theory is expressed in mathematical notation as:

$$B = f(P, E)$$

where behaviour (B) is a function (f) of the person (P) and the environment (E)

By defining the psychological environment as a *field*, Lewin established a construct and method upon which the relationship between person (P) and environment (E) could be expressed, and thus, studied (Cartwright 1952). Later, Lawton (1980) argues that the person-environment (P-E) relationship imposes on the adaptation level of individual competence, and modified the formula, thus: $B = f$ (P, E, P-E). Lawton also developed the Environmental Docility Hypothesis (Lawton and Simon 1968) that expresses the vulnerability of less competent individuals in functioning when the press exceeds competence. The immense contributions of Lawton to environmental gerontology were his deep insights in articulating the supportive and compensative roles that the environment can play towards enhancing the quality of life of older adults. Embedded in them is the intrinsic potential for every person to adapt to and nurture requisite competence and resilience in overcoming challenges. Individual abilities vary among attributes such as physical, mental and cognitive health, as well as emotional or affective functioning, all of which forming the basic constitution of personhood. Furthermore, quality of life skillsets comprising ability to plan, sense of efficacy, autodidactic ability to learn, mastery and resourcefulness, are quintessential to individual competence. Invariably, the environment plays a nurturing role if well-conceived, or conversely, become an impediment to wellbeing.

Putting into perspective the multifaceted dimensions of ageing in a complex urban environment, the postulation of this essay is that psychosocial-urban resilience is a collective behaviour rather than merely an individual ability to adapt (Fig. 1). Beside the P-E factor identified by Lawton, the behavioural trait of resilience is engendered in deep ties forged over time through shared values, activities and spatial experiences. This relationship between different active integral components that shape resilience could be expressed in an extended formula: $B_R = f$(P-E-C_P-A-T-I) where,

B_R = Resilience (psychosocial-urban)
P = Person (individual or group)
E = Environment (psychosocial and geospatial)
C_P = Commons (multi-agent community–P_1, P_2, $P_3 \ldots P_n$)
A = Activities (collective relationships)
T = Time (duration of collective relationships)
I = Intensity (intensity of collective relationships)

Alluding to Lewin's field theory (Cartwright 1952), the interconnected and interdependent relationship of multiple factors in $B_R = f$(P-E-C_P-A-T-I) is dynamic in nature. In this comprehensive formulation of resilience as a human trait involving collective adaptive abilities, the key variables extend beyond Lawton's factors to include

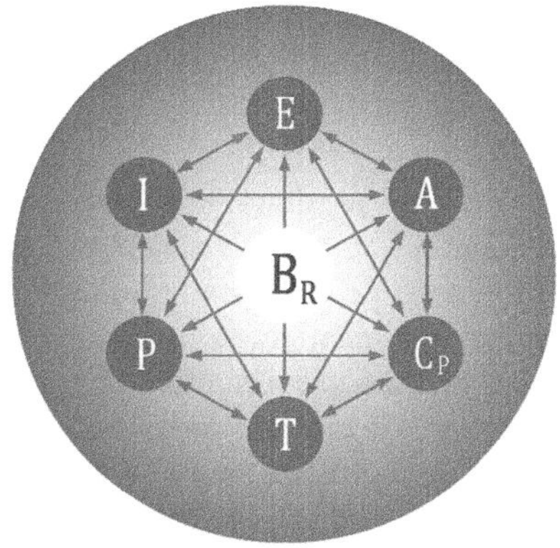

Fig. 1 The field theory model of resilience. *Source* Author

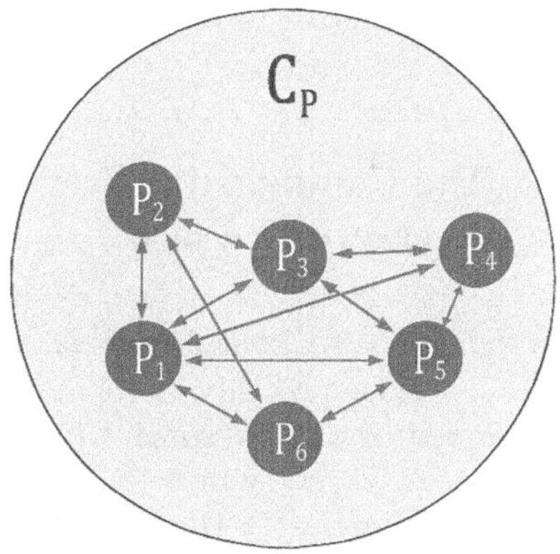

Fig. 2 Multi-agent interrelationship of commons. *Source* Author

a multi-agent *commons* (C_P)—the community—taken as a whole beyond individual or group experiences (Fig. 2). Resilience at the agency level is crucial to individual wellbeing through adaptation, but does not necessarily contribute to inter-agent cohesion, which is a prerequisite for resiliency of a community. A collective psyche with shared experiences and values is quintessential when upscaling resilience to a larger psychosocial-urban construct, like a neighbourhood. This idea of commons aligns with Gestalt psychology, as a whole, shared experience beyond and different from the mere sum of individual behaviours in a community. Community interaction begets communality, sometimes depicted as a herd or tribal behaviour.

Moreover, by adding a collective multi-agent actor, psychosocial-urban resilience is necessarily constructed from shared activities (A) over time (T), the intensity (I) of which affects how deeply the psychosocial bonds are emerged, embedded, reinforced through repetition, and recalled as collective memories. Here, the environment (E) provides an omnipresent, ubiquitous and ineradicable geospatial context for psychosocial-urban resilience to be forged. For this reason, a virtual community lacks cognisability of tangible materiality of space to produce deep collective memories. In this instance of interconnectedness, Lawton's P-E transaction is absorbed in a multifactor dynamic that includes the person-environment relationship. The environment not only acts on the person but also the commons, as it also does on activities, time and intensity.

Resilience is a counterforce to sudden and unpredictable acute shocks, but also an antidote to slow and known chronic stresses—different orders of magnitude. Undoubtedly, the active agent in fostering resilience is individual ability to adapt and thrive in difficult circumstances of people (psychosocial resilience), phenomenon (situational resilience) and place (environmental or urban resilience). In reciprocity, people, phenomenon and place offer the opportunities for each person to cultivate resilience. Place familiarity, an aspect of person-environment relationship engenders enduring bonds that support active ageing and resilient community. The key argument here is that psychosocial and urban resilience presupposes a deep collective identity acquired through familiarity in a convivial and supportive environment over protracted time.

4 The Commons: A Prerequisite for Community Ageing-with-Place

Although the socio-ecological perspective focuses on resilience at the macro-level, its qualities are equally, if not more appropriate, at the meso-level such as a neighbourhood. The *100 Resilient Cities* programme defines resilience as "the ability of a system, entity, community or person to withstand shocks while still maintaining its essential functions" (The Rockefeller Foundation 2015, p. 3). It lists seven qualities of resilient cities as: reflectiveness, resourcefulness, robustness, redundancy, flexibility, inclusiveness and integration. Resilience is deemed a complex urban systems property (Resilience Alliance 2010) based on adaptation and transformation over a long term (Chelleri and Olazabal 2012). The high-density urban neighbourhood community is one such complex system operating at the meso-level.

What constitutes community is a contested notion attributed by scholars to a tectonic shift in the mid-1980s known as the 'cultural turn' (Delanty 2010). No longer deemed as merely a social practice but rather for its symbolic structure where community is what people define it to be (Cohen 1985), this social constructionist view has influenced discussions in recent decades. Essentially, it is a change in perspective from geographically based social interaction to concerns with symbolic meanings and identity, including place identity and perception of boundary (Barth 1965). Anderson's (1983) postulation of imagined rather than lived spatial constructs extends this to cognitive and symbolic structures. Against this tide, critics argued that the cultural turn has resulted in a loss of interest in social relations (Amit 2002) and that community involves both an idea and a real social phenomenon (Delanty 2002). The evolving definitions and understanding of community is ever more ineluctable in today's increasing omnipresence of the virtual realm fuelled by smart technologies. Prevailing concerns focus on the social deficit engendered by an enchantment with technologies in modern and affluent cities, which in reality promotes solitude despite promising instantaneous connectivity (Montgomery 2013).

Equally, ageing-in-place is well argued by both proponents and detractors advancing different perspectives covering various knowledge areas like place theory, health and developmental sciences, social and environmental gerontology. Scholars define ageing-in-place through lenses of functional, symbolic and emotional filtering of what home, neighbourhood and community mean (Wiles et al. 2011, 2012); how it relates to housing and eldercare options (Bayer and Harper 2000; Judd et al. 2010; Howden-Chapman et al. 1999; Means 2007); access to health and support services (Howden-Chapman et al. 1999; Lawler 2001); and importance of neighbourhoods and communities in offering different opportunities including post-retirement employment (Oswald et al. 2010; Mak and Woo 2013; Pleau and Shauman 2013; Sonnet et al. 2014; Walker 2006).

There is general alignment between advocacy of ageing-in-place and the notion of *active ageing* (WHO 2007). However, interventions in physical and social environments are needed in order to achieve the desired outcome of older adults maintaining autonomy within a community (Buffel et al. 2012). Research indicates the strong relationship between satisfaction with residential environment and quality of life (Perez et al. 2001); ageing-in-place is thus seen as a cost-effective solution with benefits for the elderly (Sixsmith and Sixsmith 2008).

It is established that ageing-in-place incurs deep implications in spatial and programmatic planning of urban environments, particularly at the neighbourhood level where everyday activities occur. What is less evident is that ageing-in-place requires not just evolvement of individual elderly persons but also the community as commons, and integration of place to inter-grow in a holistic, congruent and organic manner. In reality, the spatial-temporal continuum is neither distinguishable nor reducible to discrete elements but rather acts as a singular entity. Therefore, a more accurate and complete depiction of the concept is *community ageing-with-place*. In this enhanced version of the common term *ageing-in-place*, the physical environment and its concomitant effects—psychosocial and environmental behaviours, perceptual and cognitive experiences—transform in tandem with a community over time. Neighbourhood place qualities, particularly familiarity, are intrinsic to population ageing in a community over time. The intricate and complex connections between place and psychosocial and environmental behaviours (Lawton and Nahemow 1973; Lawton and Simon 1968), as well as perceptual-cognitive experiences (Kaplan and Kaplan 1983) are well argued.

5 Place Familiarity: The Asset of an Active Ageing Community

Each person is a locus of many human groups in urban living, inhabiting different spatial environments over a lifetime (Fung 2015). Accumulating memories and impressions of places experienced, every individual has an environmental autobiography (Israel 2003), a treasure chest of poetic moments associated with space and place (Bachelard 1964). Our memories and attachment to place vary as a child develops from an initial autocentric understanding of reality in early childhood, to an allocentric view of the world (Schachtel 1959). This transition entails familiarity, continuity and assurance from constancy of the environment.

The corollary of a deficit in place familiarity is disaffection with place and social identity, reduced sense of safety and security, and diminished memory and meaning. Since psychosocial and urban resilience is closely associated with these qualities, place familiarity is a requisite condition of resilience in all its various expressions. To appreciate human development, it is necessary to understand how the factors of person, place and time combine to influence behaviours and change (Lerner 2007; Diaz Moore 2014). The ability of individuals to withstand unanticipated shocks as well as known stresses is invariably linked to the cohesive interdependence of place, community and self, with time as the factor of change. Without such a solid foundation of interconnections, resilience, be it individual, situational, social or environmental, would be difficult to achieve.

Place familiarity is closely associated with, and deeply impacts on, the formation of psychosocial-urban resilience in the neighbourhood. Not only does place familiarity foster social and spatial identity—the underlying premise of place attachment—but it also serves as a crucible for forging a deep sense of home and refuge, including symbolic roles recalled in times of crisis and immense stress. For seniors, the unassuming assurance of a familiar neighbourhood is pivotal to encouraging out-of-home activities for daily routines, thus averting social isolation.

Place integration—the precursor of familiarity, identity, attachment and memory—is to become at-home by interacting with people and environment, in which habits, imagination and habitual thoughts inform an ongoing process (Cutchin 2013). The formation of place relationships over time involves individual perception, and is central to happiness and wellbeing. Many nested layers, including sociological, psychological, anthropological and geographical overlay our notion of home (Rowles and Chaudhury 2005). Since antiquity, geospatial effects are key to human experience and the understanding of place as phenomenology of the everyday world. Home is much more than the physical locale of habitat, but a place to dwell in the sense of Heidegger's *dasein* or *being* in the world, and thus is full of attachments and meanings. To integrate with a place means to be fully attuned to the significance of everyday life in all its nuances.

Attachment to familiar places at home base is high for older persons as it provides a crucial link to the active phase of adult life (Hidalgo and Hernandez 2001), and symbolises a phase of self-achievement in material possessions and circumstances (Antonelli et al. 2000; Lawton 1990). The importance of growing old at the home of one's adult phase is crucial to most elderly persons due to familiarity (Wahl 2001), and the home evokes the strongest person-environment link in older people (Wahl et al. 1999). This accounts for the general reluctance of most elderly people towards institutionalisation in a collective care residential environment. Conversely, oftentimes the decision to relocate is dependent on an elderly's ability to continue functioning independently at home in performing activities of daily living.

Ironically, the sense of home and place is most poignant when a person has to be displaced and transferred to another locale, a common occurrence in city living but especially contemporary technology-empowered urban nomadic lifestyles. Older adults, who relocate to nursing homes, retirement villages, assisted living or similar residential aged care facilities, are highly susceptible to uprooting and dislocation. Many elderly people monetise their matrimonial home and move to a smaller apartment in order to recoup some monies at old age or due to empty nesting where the house becomes too large after their children marry and move out. Such transitions are deemed essential and integral to place experience and identity (Buttimer 1980; Seamon 1979), place transitions by juxtaposition help to shape a person's perception of home, which is a lifetime occurrence (Watkins and Hosier 2005). Although presence of the absent might play a role in reinforcing a sense of home and place, the fact that it is achieved after a loss of connection to both, makes it of little practical purpose except to reside in memory. In effect, it poses a problem for people with dementia who could only recollect their earlier home place, exacerbating wandering behaviour in an unfamiliar environment like a nursing home or retirement village. Moreover, under such constant shifts, how are the psychosocial constructs and modalities of experience (Lawton 1998; Weisman et al. 2000) affected? It is known that people's cognitive-affective appraisals could change quickly over a day in response to changing environmental conditions, such as fluctuating daylight (Diaz Moore 2014). Hence, the aggregated daily experiences of place over a life-course are needed to offer stability and constancy of reference, fostering resilience.

With decline in physiological and cognitive abilities from ageing, the relationship with place manifesting as place identity and place attachment become more significant (Rubinstein and Parmalee 1992). Any transfer to a new place would require much of those complex relationships to be re-established or re-negotiated, including reshaping former habits of place (Rowles and Watkins 2003). The process involves not only new physical settings but also psychosocial and cognitive re-orientation. Ageing-with-place should equally concern place transformation, and its reciprocal effect in shaping wellbeing (Cutchin 2013). Where a well-planned place retains substantially its essential character over time, older adults who grew up in that familiar setting are naturally better adapted to function well. Conversely, constant erasure of known features in the neighbourhood wreaks disorientation and bewilderment from a lack of recognisability.

The sociocultural environment becomes important with increasing frailty and diminished mental, physical and cognitive abilities at old age, supporting activities of daily living in the community. Such intersections between materiality and everyday routines help to form enduring habits or *habitus* (Bourdieu 1977), manifesting as foundational metapatterns of human-environment relationship at individual and communal levels. Community ageing-with-place is arguably the most sustainable way to provide long-term care for the aged and is closely associated with forging of psychosocial-urban resilience. Neighbourhood familiarity is the key to compensating for diminished abilities by offering constancy in place and social relationships.

This practical notion of place integration is founded on the philosophy of John Dewey, who argues that continuity of person and place is crucial to the person-environment fit as a form of inquiry. Under the ambit of geographical pragmatism (Cutchin 2013), scholars see this transactional mode as a part of forming positive collaborations between people, as well as people and environment.

6 Designing for Ageing in High-Density Urban Neighbourhoods

Architectural and urban theories constantly engage the complex human-environment relationships across layers that include function, spatial perception, sensory experience and symbolic meaning. For Amos Rapoport, the environment is not a mere assemblage of people and things but reflects complex interrelationships between them through a process of enculturation and values formation (Rapoport 1969). Enduring human relationships have an underlying order manifesting as pattern and structure of a human settlement, leading Christopher Alexander to postulate the inverse reciprocity that the desired urban experience and spatial qualities could be attained through reconstruction by understanding and emulating the built patterns of strong appeal (Alexander et al. 1977). The phenomenologist Christian Norberg-Schulz in his seminal book *Genius Loci* went further to argue for the existence of the *spirit of place* in emergence of spatial-temporal phenomena (Norberg-Schulz 1980).

In his treatise *City Planning according to Artistic Principles,* Sitte (1889) critiques gridiron planning as the output of mechanical thinking, lacking in sensitivity and imagination. Furthermore, many urban theorists advocate organically emerged urban qualities like *eyes-upon-the-street* natural surveillance of street-based communities (Jacobs 1962), high imageability of place (Lynch 1960), and place identity and territorial ownership of defensible spaces (Newman 1972). Increasing disenchantment with Modernist planning made reviewing our understanding of good urban environment more urgent.

Moreover, the incessant overlaying of ceaseless stimuli in dense urban environments results in extreme stress build-up, diametrically opposite of an idealised picturesque tranquillity. Genteel neighbourhoods of conviviality seem alien and antithetical to high-density urban living, specifically if structured and designed according to contrived top-down Modernist planning principles. Devoid of the organic emergence of meanings from small-scale bottom-up initiatives by the collective commons, how could contemporary planned urban neighbourhoods ever come close to fostering conviviality?

Spatial morphology studies on ageing tend to focus on walkability and accessibility to amenities, specifically in attempts to introduce quantitative measurements to adduce evidence. Yet, this merely continues the reductive understanding of neighbourhood planning and design. Moreover, such works do not examine the actual challenges of wayfinding faced by the elderly, particularly the experience of place.

It assumes that physical connectivity alone would suffice as a proxy of good accessibility and meaningful spatial experience. Complicating this is the prevailing enchantment with smart technology as the panacea in enabling human relationships. In high-rise, high-density urban conditions, spatial complexities evince deep challenges for the elderly, particularly for those afflicted by dementia or cognitive impairment. This requires a high-touch, rather than high-tech, approach towards supportive real-world experiences.

Many current studies on urban design of age-friendly environments focus on landuse and functionality such as walkability, accessibility, connectivity, mobility (first- and last-mile commuting) and service distances. The deployment of geographic information system (GIS) and other geo-locational technologies for big data analytics lends an air of authority to such quantitative studies. In spite of its fine-granularity of data, the enriched and dense information that matters most to older adults' actual psychosocial experiences of the neighbourhood can never be captured in all its nuances. The process of transforming *space* from a mere physical container of human activities into a collectively identified *place* is a complex endeavour that imbues diverse meanings over time. The attributes of place-making were well discussed in architectural and urban theories, albeit, developed mostly in low-density situations, including but not limited to creating a sense of familiarity, centring of an individual's place in the world, affording security and safety, ownership and control, territoriality, psychometric comfort, display and identity (Rowles and Bernard 2013).

Rigorous research of high-density urban environments is needed to provide a holistic and integrated understanding of the nuances of place-making conducive to ageing. More importantly, new innovative typologies of high-density urban neighbourhoods must emerge that are truly supportive and compensative of aged care in an active integral manner, rather than merely reactive to instances of deficient planning or disconnected design features. Tremendous benefits could be gained for studying high-density Asian cities like Singapore, Hong Kong and Seoul—where large populations reside in high-rise public housing—so that the consequential repercussions can be highlighted for many other cities that seek to emulate them in Asia and beyond.

7 Singapore: An Illustration of Design Challenges

Singapore is renowned for her high-rise, high-density public housing programme where more than 80% of its population reside in such residential developments by the Housing and Development Board (HDB). Some argue that its success lies mainly in organisational planning and judicious management of functional needs. Although efforts were made to instil identity, primarily through landscaping and cosmetic use of design motifs and colours, most of its housing estates spot a visual homogeneity that led Koolhaas and Mau (1995) to provocatively pronounce Singapore as a *generic city* built through *tabula rasa* top-down planning. Critiques of the HDB programme highlight its banal environment, car-centric planning, lack of distinct place identity, and deficit of deep community spiritedness—views that are contested by others.

The constant lamentations expressed in public media about the loss of community spirit and mutual self-help lend weight to such criticisms. Notwithstanding the divergent views, a rapidly ageing population requires urgent re-examination and rethinking of how HDB neighbourhoods could better promote a sense of rootedness, through the spatial construct of place familiarity.

The crux of challenges in designing high-density age-friendly neighbourhoods to attain psychosocial-urban resilience lies in close integration of planning, urban design, programming, and community engagement by entrenching architecture and place-making as active integral components of public policies. For instance, the current top-down method in developing HDB's Build-to-Order (BTO) apartments would benefit immensely from greater public participation in upstream processes through co-creation such as participatory design and focus-group discussion. In fostering a sense of ownership, there is no substitute to partaking in shaping one's neighbourhood from the outset. This top-down combined with bottom-up approach would require adjustments to current public policies and bureaucratic thinking.

Furthermore, the actual design of neighbourhoods must prioritise human-centred needs over other considerations, although operational practicalities need not be entirely negated. Rather than maximising profit, the preferred attitude would be one of optimisation towards overall enhancement of community wellbeing. A case in point: the need to create a child-and-elderly safe zone to encourage high usage of urban social spaces and facilities implies that vehicular movement within the neighbourhood should be made secondary or even eradicated altogether. This entails car-lite or car-free living, augmented by good infrastructure for personal mobility options e.g. walking, cycling, motorised devices, etc.—a daunting goal given the legacy of a car-loving society like Singapore. Car-centric planning tends to disrupt the creation of convivial social spaces for community interaction and bonding (Fig. 3).

Another instance is community gardens that sprout organically in the open grass patches beneath HDB housing blocks. Paradoxically, these are mostly fenced-in despite its ground-up origin, requiring a temporary use licence, and not readily accessible to the community. The irony is even more poignant when juxtaposed against open allotment gardens in public parks, located away from housing neighbourhoods, which charge a monthly rental per plot. Should not community gardening, as a shared activity that nurtures psychosocial resilience be available free, openly accessible and entirely owned and managed by the community? This ambiguity in spatial programming is exacerbated by ornamental landscapes commonly found in HDB neighbourhoods created through a top-down process as an aesthetic element, manicured and managed by the town council instead of residents (Figs. 4, 5 and 6). Such contrived features create a detached identification rather than deep place identity as when the commons are co-designed and co-created by residents. Promoting social interaction to cultivate collective resilience involves trade-offs to attain urban commons that not only support functional but also psychosocial needs of place identity and attachment.

For older adults whose sensory abilities are diminished, daily routines like wayfinding and safe passage within the neighbourhood are crucial in maintaining out-of-home activities. Many elderly prefer to stay home to avoid falling and its related injuries, as older people need a longer period for physical recovery.

Fig. 3 Car-centric planning in housing estates. *Source* Author

The adverse consequence is social isolation, resulting in health deterioration and reduced quality of life. A well-designed neighbourhood should facilitate intuitive navigation, safe access and seamless connection to amenities. This promotes frequent usage and purposeful engagement with the neighbourhood spaces and community, reinforcing place integration. Place familiarity means having a cognitive map or model of an environment in memory, which facilitates wayfinding functions that are less dependent on direct information from the environment (Kaplan and Kaplan 1983).

8 Planning and Design Considerations

Whilst in principle the attributes of good city planning, place-making and urban design—mostly conceived in low-rise, low-density cities of the developed west—are apparently relevant, operationalising such ideals differ from city to city, country to country. First, the *morphological scale and spatial complexities of densification* pose a challenge to many desired attributes. These include good wayfinding, privacy, sense of human scale, visual sightlines, access to nature, connectivity, accessibility, vertical circulation, integrated multiuse functions, place identity, etc. For elderly

Fig. 4 A fenced-in community garden. *Source* Author

urbanites, deficient design could impede free movement in the neighbourhood or city. Secondly, *contextual and cultural differences* mean that the desirability of attributes is necessarily dissimilar. Values, symbolism and meanings associated with certain aspects of western culture may not translate equally for non-western societies. An oft-cited instance is the neoliberal west's emphasis on individuality that juxtaposes against Asian sociocultural preferences for the collective. Filial piety and respect for the elderly are still practised in Asian societies, which affect sociocultural perception of the commons. Thirdly, *national policies and programmes* differ greatly depending on the socio-political ideology of the government in place. Ageing-related national social and healthcare programmes, as well as their financing, are widely diverse and different between cities and countries.

From the foregoing discussion, the following considerations may be surmised when planning and designing urban neighbourhoods for collective resilience and community ageing-with-place (Fig. 7).

9 Future Urban Neighbourhoods

The foregoing discussion illuminates the importance of place familiarity in forging resilience, through fostering strong social relationships in urban neighbourhoods.

Fig. 5 Allotment garden in a public park. *Source* Author

It advances the idea of *community ageing-with-place* as the interdependent and concomitant flourishing of individuals, social bonds and place-making. Here lies the imperative for future neighbourhoods to attain urban and psychosocial resilience, which is crucial for an ageing society.

Interdisciplinary studies of urban design and ageing are on the rise worldwide. The broadening interest in environmental gerontology was attributed to a *spatial turn*—equivalent to what is commonly called the *cultural turn* in the social sciences. Furthermore, global phenomena such as unprecedented population ageing, emerging healthcare models and other impactful changes engendered practical and academic interests in space and place (Andrews and Phillips 2005). The underlying argument here is that many benefits can accrue if the environment and community at the neighbourhood level are supportive and compensative for its older residents. This aligns with current global interest in a community-based approach towards the sustainability of long-term care for rapid population ageing. Attaining urban and psychosocial resilience also accords well with the global vision articulated by the *New Urban Agenda* (UN 2017), which acts upon the United Nations' *2030 Agenda for Sustainable Development* (UN 2015).

Against the backdrop of prevailing theories, models and professional practice, what might future urban neighbourhoods be like as a place conducive for fostering resilient communities that age-with-place? In attempting to discuss this question, a

Fig. 6 Ornamental landscape design. *Source* Author

Planning and Design Considerations		Remarks
1	A systems approach to the commons is requisite in fostering collective resilience and community ageing-with-place	Organised holistic actions offer systemic benefits that are not available to the individual
2	The urban neighbourhood acts as a *field* that impinges on each elderly resident, and is itself affected by the collective actions of its constituents	Physical environment affects people as a non-reductive continuum
3	A congruent and comprehensible neighbourhood design aids elderly residents in perception-cognition of place, and reinforces familiarity, identity, memory and attachment	Enabling elderly people to intuitively understand their environment improves their quality of life
4	Integration of daily activities and social spaces helps to nurture individual resilience of elderly residents and collective resilience of the community	Routines engender familiarity for both the individual and the collective
5	Community participation in co-creating shared commons engenders psychosocial and urban resilience for elderly residents	Shared activities promote identification and ownership of the commons
6	Familiarity arises from constancy and routine access of place, people and phenomena (activities), promoting out-of-home activities of elderly residents and averting social isolation	Regularity and recurring interactions enhance social bonding
7	Time is the cohesive agentñ space, time and activities should be planned together	Integrated space, time and activities reinforces the person-environment fit
8	Place familiarity is affected if pre-cognised aspects of the environment are substantially changed or erased, resulting in geospatial disorientation in wayfinding	Constancy of environment helps in maintaining recognition, association and sense of belonging

Fig. 7 Planning and design principles for resilient urban neighbourhoods. *Source* Author

useful starting point is to understand how global trends are evolving and examine their repercussions on ageing, place familiarity and fostering of resilience.

Concurrent with global rapid ageing populations, the following trends in urban Asia merit attention: (i) an increase in elderly people living alone in single-person households, (ii) rising expectation for diverse lifestyles from baby-boomer and younger generations, (iii) changes in family structure with reduction in dependency support for older adults, (iv) demand for more and enriched options of long-term care, and (v) urban live-and-work displacement, and (vi) the advent of disruptive technologies, to name a few prominent ones. Together these current and emerging trends coalesce to make the re-imagination of future urban neighbourhoods a critical and urgent task. In high-density urban settings, the commons are further threatened by crowdedness, heightened environmental stresses, and increased mobility and safety hazards—challenges that are exacerbated by the diminished abilities of older adults. Although population and spatial densities offer advantages of natural surveillance (Jacobs 1962) and convenient access to services, there is a trade-off in urban qualities, such as liveability and conviviality when density exceeds an optimal point. Judicious integration of people, place and programmes is needed to offset the adverse impact of endless upscaling of human interaction in a congested environment.

A matter of pressing concern lies in defining the optimal and appropriate size of an age-friendly neighbourhood that could support a good range of services without imposing excessive travel, overwhelming stresses from densification or diluting opportunities of community cultivation. Walkability is well studied but this preoccupation with accessibility and mobility needs to be juxtaposed against other considerations, such as appropriate population size and mix, scale of buildings and open spaces, spatial morphology, wayfinding, social structure and organisation, etc. Significantly, neighbourhood size affects the nurturing of place identity and familiarity as well as deep social ties and collective resilience. Disparity in interpretations of a neighbourhood community is evident in the huge difference between what Clarence Perry conceived as ideally 4800 to 9000 residents over 65 ha of land (Perry 1929), which is basically low-density, and actual residential conditions in highly compact Asian cities like Hong Kong, Seoul, Mumbai, Shanghai, Shenzhen, Tokyo and Singapore. These cities were built on much higher densities—a trend that is continuing to increase, as urban living is made more appealing through better urban design.

It is inevitable that disparity exists between an ideal human-scaled development and actual realities in high-density living. Repercussions of high-density include imposing spatial scale and overwhelming experiences, environmental impact (like urban heat island effect), constricted wind flow and daylight, as well as alienation and difficulties in forming or nurturing social ties, impacting psychosocial-urban resilience. How could a deep community ever be formed under such asymmetry of spatial quality, a far cry from the conviviality of genteel low-density environments? In making future urban neighbourhoods as places that foster strong psychosocial-urban resilience through familiarity and identity, immense will is needed to embed from the outset, planning and urban design principles that promote intuitive support for older residents, natural surveillance by the community, a sense of shared ownership of spaces and places—collectively the commons, while seamlessly deploying smart technologies to augment the analogue spatial-sensorial experiences.

Human needs have evolved over time, are multifaceted and difficult to define. Maslow (1970) offers a simple framework that encapsulates different human needs —often cited by others as a *hierarchy of needs*—and how we cope with challenges. He also advocates a holistic-dynamic point of view to counter prevailing atomistic, taxonomic, static, causal and mechanical worldviews. Moreover, Maslow postulates that self-actualising people grow through motivation, the self-ability that ensures stability in the face of adversity, relatively independent of the physical and social environment—a point of contention. He concluded that "while a good environment fosters good personalities, this relationship is far from perfect, and furthermore, the definition of *good* (sic) environment has to change markedly to stress spiritual and psychological as well as material and economic forces" (Maslow 1970, p. 279). One surmises from Maslow that appropriate environments could foster resilience and allow a person to thrive despite constantly shifting cause-effect relationships.

The larger concern remains on whether in the relentless pursuit of landuse intensification in the name of high-density planning as in the case of Singapore, we sacrifice the potential for cultivating good community formation that is quintessential to deep social resilience. Should a new city planning paradigm that favours optimisation, spatially, environmentally and socially, replace the current maxim of maximisation? What trade-offs should we accept between total exploitation of natural resources and attaining psychosocial-urban resilience through place familiarity? These emerging questions beg urgent research and new knowledge of the human condition that would contribute to true and lasting wisdom.

References

Alexander C, Ishikawa S, Silverstein M (1977) A pattern language: towns, buildings, construction. Oxford University Press

Amit V (2002) Realizing community: concepts, social relationships and sentiments. Routledge

Anderson B (1983) Imaginary communities: reflections on the origin and spread of nationalism. Verso

Andrews GJ, Phillips DR (2005) Ageing and place: perspectives, policy, practice. Routledge

Antonelli E, Rubini V, Fassone C (2000) The self-concept in institutionalized and non-institutionalized elderly person. J Environ Psychol 20:151–164

Antonovsky A (1979) Health, stress and coping. Jossey-Bass

Bachelard G (1964) The poetics of space. Beacon Press

Baltes PB, Baltes MM (1990) Successful aging: perspectives from the behavioral sciences. Cambridge University Press

Barth F (ed) (1965) Ethnic groups and boundaries. Allen & Unwin

Bayer AH, Harper L (2000) Fixing to stay: a national survey of housing and home modification issues. AARP

Blackburn E, Epel E (2017) The telomere effect: a revolutionary approach to living younger, healthier, longer. Orion Spring

Bourdieu P (1977) Outline of a theory of practice. CUP

Buffel T, Phillipson C, Scharf T (2012) Ageing in urban environments: developing 'age-friendly' cities. Crit Soc Policy 32(4):597–617

Buttimer A (1980) Home, reach, and the sense of place. The Human Experience of Space and Place 3:166–187
Cartwright D (ed) (1952) Field theory in social science: selected theoretical papers by Kurt Lewin. Social Science Paperbacks
Chelleri L, Olazabal M (eds) (2012) Multidisciplinary perspectives on urban resilience: a workshop report. BC3 Basque Centre for Climate Change
Cohen A (1985) The symbolic construction of community. Tavistock
Cutchin MP (2013) The complex process of becoming at-home in assisted living. In: Rowles GD, Bernard M (eds) Environmental gerontology: making meaningful places in old age. Springer Publishing Company
DeBaggio T (2003) When it gets dark: an enlightened reflection on life with Alzheimer's. The Free Press
Delanty G (2002) Communitarianism and citizenship. In: Isin EF, Turner BF (eds) Handbook of citizenship studies. Sage, pp 161–174
Delanty G (2010) Community, 2nd edn. Routledge
Diaz Moore K (2014) An ecological framework of place: situating environmental gerontology within a life course perspective. Int J Aging Hum Dev 79(3):183–209
Fung JC (2015) Dementia design sourcebook. National University of Singapore, Department of Architecture
Garmerzy N (1973) Competence and adaptation in adult schizophrenic patients and children at risk. In: Deen SR (ed) Schizophrenia: the first ten dean award lectures. MSS Information, pp 163–204
Gesler W (1992) Therapeutic landscapes: medical issues in light of the new cultural geography. Soc Sci Med 34:735–746
Harris PB (ed) (2002) The person with Alzheimer's disease: pathways to understanding the experience. The John Hopkins University Press
Harris PB (2008) Another wrinkle in the debate about successful ageing: the undervalued concept of resilience and the lived experience of dementia. Int J Aging Hum Dev 67(1):43–61
Hidalgo MC, Hernandez B (2001) Place attachment: conceptual and empirical questions. J Environ Psychol 21:273–281
Holstein MB, Minkler M (2003) Self, society, and the "new gerontology". Gerontol 43:787–796
Howden-Chapman P, Crane J, Signal L (1999) Housing and health in older people: ageing in place. Soc Policy J N Z 13:14
Israel T (2003) Some place like home: using design psychology to design ideal places. Wiley+Academy
Jacobs J (1962) The death and life of great American cities. Jonathan Cape
Judd B, Olsberg D, Quinn J, Groenhart L, Demirbilek O (2010) Dwelling, land and neighbourhood use by older home owners. AHURI Final Report No. 144. Australian Housing and Urban Research Institute
Kahana E, Kahana B (1996) Conceptual and empirical advances in understanding aging well through protective adaptation. In: Bengtson V (ed) Adulthood and aging: research on continuities and discontinuities. Springer, pp 18–41
Kaplan S, Kaplan R (1983) Cognition and environment: functioning in an uncertain world. Ulrich's Bookstore
Kitwood T (1997) Dementia reconsidered: the person comes first. Open University Press
Koolhaas R, Mau B (1995) S,M,L,XL: small, medium, large, extra-large. The Monacelli Press, Inc.
Lawler EJ (2001) An affect theory of social exchange. Am J Sociol 107(2):321–352
Lawton MP (1980) Environment and aging. Brooks/Cole Publishing Company
Lawton MP (1990) Residential environment and self-directedness among older people. Am Psychol 45(5):638–640
Lawton MP (1998) Environment and aging: theory revisited. In: Scheidt RJ, Windley PG (eds) Environment and aging theory: a focus on housing. Greenwood Press, pp 1–31
Lawton MP (1999) Environmental taxonomy: generalizations from research with older adults. In: Friedman S, Wachs T (eds) Measuring environment across the life span: emerging methods and concepts. American Psychological Association, Washington, D.C., pp 91–124

Lawton MP, Nahemow L (1973) Ecology and the aging process. In: Eisdorfer C, Lawton MP (eds) The psychology of adult development and aging. American Psychological Association, pp 619–674
Lawton MP, Simon B (1968) The ecology of solid relationships in housing for the elderly. Gerontologist 8:108–115
Lerner R (2007) Developmental science, developmental systems, and contemporary theories of human development. In: Damon W, Lerner R (eds) Handbook of child psychology, vol 1. Wiley, pp 1–17
Lewin K, Lorsch K (1939) Mathematical constructs in psychology and sociology. J Unified Sci 9:397–403
Luthar SS, Zelazo LB (2003) Research on resilience. In: Luthar SS (ed) Reliance and vulnerability: adaptation in the context of childhood adversity. Cambridge University Press, pp 510–549
Lynch K (1960) The image of the city. The MIT Press
Lyman K (1989) Bringing the social back in: a critique of the biomedicalization of dementia. Gerontol 29:597–605
Mak B, Woo J (2013) Retirement and postretirement issues. In: Woo J (ed) Aging in Hong Kong: a comparative perspective. Springer
Maslow AH (1970) Motivation and personality, 2nd ed. Harper & Row
Masten AS (2001) Ordinary magic: resilience processes in development. Am Psychol 56:227–238
Masten AS, Coatsworth JD (1998) The development of competence in favorable and unfavorable environments: lessons from successful children. Am Psychol 53:205–222
Masten AS, Powell JL (2003) A resilience framework for research, policy, and practice. In: Luthar SS (ed) Reliance and vulnerability: adaptation in the context of childhood adversity. Cambridge University Press, pp 1–25
Means R (2007) Safe as houses? Ageing in place and vulnerable older people in the UK. Soc Policy Adm 41(1):65–85
Montgomery C (2013) Happy city: transforming our lives through urban design. Farrar, Straus and Girox
Moody HR (2005) From successful aging to conscious aging. In: Wylke MC, Whilehouse PJ, Morris DC (eds) Successful aging through the life span. Springer
Newman O (1972) Defensible space: people and design in the violent city. Macmillan
Norberg-Schulz C (1980) Genius loci: towards a phenomenology of architecture. Academy Editions
Oswald F, Jopp D, Rott C, Wahl HW (2010) Is aging in place a resource for or risk to life satisfaction? Gerontol 51(2):238–250
Perez FR, Fernandez-Mayoralas G, Fernandez EPR, Abuin JMR (2001) Ageing in place: predictors of the residential satisfaction of elderly. Soc Indic Res 54:173–208
Perry CA (1929) The neighborhood unit. In: Neighborhood and community planning. Regional Plan of New York and its Environs, pp 22–132
Pleau R, Shauman K (2013) Trends and correlates of post-retirement employment, 1977–2009. Hum Relat 66(1):113–141
Poulton R, Moffitt TE, Silva P (2015) The Dunedin multidisciplinary health and development study: overview of the first 40 years, with an eye to the future. Soc Psychiatry Psychiatr Epidemiol 50(5):679–693
Rapoport A (1969) House, form and culture. Prentice-Hall
Resilience Alliance (2010) Assessing resilience in social-ecological systems: workbook for practitioners (Version 2.0)
Riley MW (1998) Letter to the editor. Gerontol 38:151
Rowe JW, Kahn RI (1997) Successful aging. Gerontol 37:433–480
Rowe JW, Kahn RI (1998) Successful aging: the MacArthur foundation study. Pantheon
Rowles GD, Bernard M (2013) The meaning and significance of place in old age. In: Rowles GD, Bernard, M (eds) Environmental gerontology: making meaningful places in old age. Springer
Rowles GD, Chaudhury H (eds) (2005) Home and identity in late life: international perspectives. Springer

Rowles GD, Watkins JF (2003) History, habit, heart, and hearth: on making spaces into places. In Schaie KW, Wahl H-W, Mollenkopf H, Oswald F (eds) Aging independently: living arrangements and mobility. Springer, pp 77–96
Rubinstein RL, Parmalee PA (1992) Attachment to place and the representation of the life course by the elderly. In: Altman I, Low S (eds) Place attachment. Plenum Press, pp 139–163
Sabat S (2001) The experience of Alzheimer's disease: life through a tangled veil. Blackwell Publishers
Schachtel E (1959) Metamorphosis: on the development of affect, perception, attention and memory. Basic Books
Seamon D (1979) Phenomenology, geography and geographical education. J Geogr High Educ 3(2):40–50
Sitte C (1889) City planning according to artistic principles. Carl Graeser, Wien. trans: Collins G, Collins CC (2006) Camillo Sitte: the birth of modern city planning. Dover
Sixsmith A, Sixsmith J (2008) Ageing in place in the United Kingdom. Ageing Int 32(3):219–235
Smith KR, Corvalán CF, Kjellström T (1999) How much global ill health is attributable to environmental factors? Epidemiology 10(5):573–584
Snowdon DA (1997) Aging and Alzheimer's disease: lessons from the Nun study. Gerontol 37(2):150–156
Snowdon DA (2003) Healthy aging and dementia: findings from the Nun study. Ann Intern Med 139(5/2):450–454
Sonnet A, Olsen H, Manfredi T (2014) Towards more inclusive ageing and employment policies: the lessons from France, the Netherlands, Norway and Switzerland. De Economist 162(4):315–339
The Rockefeller Foundation (2015) 100 resilient cities. The Rockefeller Foundation
Tornstam L (2005) Gerotranscendence: a developmental theory of positive aging. Springer
United Nations (2015) Transforming our world: the 2030 agenda for sustainable development. Retrieved in Dec 2018 from: https://sustainabledevelopment.un.org/content/documents/21252030%20Agenda%20for%20Sustainable%20Development%20web.pdf
United Nations (2017) New urban agenda. UN Habitat III Secretariat. Retrieved in Dec 2018 from: http://habitat3.org/the-new-urban-agenda/
Valiant GE (2004) Positive aging. In: Linley A, Joseph S (eds) Positive psychology in practice. Wiley, pp 521–539
Wahl HW (2001) Environmental influences on aging and behaviour. In: Birren JE, Shaie KW (eds) Handbook of the psychology of aging, 5th edn. Academic Press
Wahl HW, Schilling O, Oswald F, Heyl V (1999) Psychosocial consequences of age-related visual impairment: comparison with mobility-impaired older adults and long-term outcome. J Gerontol 54B(5):304–316
Walker A (2006) Active ageing in employment: its meaning and potential. Asia-Pac Rev 13(1):78–93
Watkins JF, Hosier AF (2005) Conceptualizing home and homelessness: a life course perspective. Home and identity in late life: international perspectives, pp 197–215
Weisman GD, Chaudhury H, Diaz Moore K (2000) Theory and practice of place: toward an integrative model. In: The many dimensions of aging: essays in honor of M. Powell Lawton. Springer, pp 3–21
Werner EE, Smith RS (1982) Vulnerable but invincible: a study of resilient children. McGraw-Hill
Wiles J, Wild K, Kepa M, Peteru C (2011) Resilient ageing in place: project recommendations & report. Faculty of Medical and Health Sciences, University of Auckland
Wiles JL, Leibing A, Guberman N, Reeve J, Allen RES (2012) The meaning of "aging in place" to older people. Gerontol 52(3):357–366
World Health Organization (WHO) (2002) Active aging: a policy framework. WHO
World Health Organization (WHO) (2007) Global age-friendly cities: a guide. WHO
Yates TM, Masten AS (2004) Fostering the future: resilience theory and the practice of positive psychology. In: Linley A, Joseph S (eds) Positive psychology in practice. Wiley, pp 521–539

Designing for Resilience in Public Housing: An Architect's Perspective

Man Kok Siew and Ivan Kurniawan Nasution

Abstract In master planning, design features can play an important role to forge strong communal bonding and resilience. Shared spaces such as common corridors, void decks, ground floor backyards, and connected walkways can act as catalysts for enhanced social participation and interactions. Fundamentally, planning must seek to bring people together and promote social cohesion so that when disasters happen, they will pull together to address emergent problems. Policy makers and planners have to adopt an integrated approach at the early planning stage to seek input from other stakeholders to respond effectively to the evolving needs of the community. They can also pay greater attention to flexibility, support the ageing community and integrate environmental features. Nursing homes for instance, can blend into urban neighbourhoods so that they become part of the residents' social life. Architects need to empathise with the diverse points of views from other disciplines and the end users, and nurture a symbiotic relationship with the different segments of the community.

Keywords Master planning · Design · Communal bonding · Resilience · Shared spaces · Flexibility

1 Introduction

My firm, MKPL Architects, has been involved in planning and designing public housing projects for the Singapore Housing and Development Board (HDB) since

---

Ivan Nasution collaborated in the writing of this paper by transcribing Siew Man Kok's perspectives and reflections.

M. K. Siew (✉)
MKPL Architects, Singapore, Singapore
e-mail: mksiew@mkpl.com.sg

I. K. Nasution
Centre for Ageing Research in the Environment, National University of Singapore, Singapore, Singapore
e-mail: sdeivan@nus.edu.sg

C.-H. Leong and L.-C. Malone-Lee (eds.), *Building Resilient Neighbourhoods in Singapore*, Advances in 21st Century Human Settlements,
https://doi.org/10.1007/978-981-13-7048-9_9

2013. Throughout my professional career, I have adhered to three main principles and objectives in our planning of public housing. They are: (a) encouraging upward social mobility; (b) maintaining social harmony and promoting social integration and, (c) creating a common shared experience. These are also the principles that will support building resilience in Singaporean urban neighbourhoods.

In December 2013, we won the design competition (it is in collaboration with Urban Design Inc.—a Canadian planning firm) to undertake the master plan for the Bidadari Housing Estate, which is the latest generation of HDB estate design, wherein the residential neighbourhoods and surrounding parks are seamlessly connected. We have also been involved in a project in the Bedok Housing Estate, where ageing is celebrated through and integrated into architecture. In this chapter, I will use these two projects to illustrate my thoughts of how we can design neighbourhoods and their communities to be more resilient. I will also recount some of my own experiences and aspirations on how we can make HDB neighbourhoods more socially cohesive through thoughtful planning and creative design.

2 Early Experiences and Inspiration

It is the positive memories of my childhood that have inspired me in my design of public housing estates and their public spaces. My happy experiences affirm my belief that neighbourhoods can bring people together and have them form bonds with one another. In design, I believe in embracing people's positive behaviour, even if sometimes it means protecting people from what they want, which may not necessarily be good for them in the long term. However, government agencies, as well as ourselves as architects and planners, are often guilty of over-doing this. We try to minimise every risk and regulate too much in our planning and design. Such an approach can only turn an otherwise warm and friendly neighbourhood into one that is cold, unimaginative and uninspiring.

I grew up in a kampong.[1] Life was different back then, in the 1960s and 70s. We were living in a multi-racial neighbourhood where everyone was close to one another. Our neighbours included a coffin shop, a barber shop, a pigsty and an assortment of other shops. My playground was literally everywhere; kids were playing at the drains and rivers or catching fish at our neighbour's pond.

When I was in primary school, my family moved from the kampong to a public housing estate in Toa Payoh, the first comprehensively planned new town in Singapore. I could still remember that it was a 3-room "improved flat" in Block 137 Lorong 1A, which no longer exists. For my family and me it was a sudden change from a ground level house to a 15th floor apartment unit (Block 126 where my brother stayed). However, maybe because I was just a kid, it was quite easy for me to adapt. I

[1] A kampong refers to a small village or community of houses, traditionally found in Singapore and other parts of South-east Asia. They are informal urban settlements. Almost all of them have been redeveloped in the course of urbanisation.

simply loved to play along with the corridors and the staircases in this new high-rise environment.

After living in Toa Payoh for twenty years, we moved to Pasir Ris, another new town. Like many Singaporean families, we "upgraded" to an executive apartment, which was quite common as families become better off with economic development of the country. I was then studying architecture, in the third year of my course at the National University of Singapore (NUS), and became acutely aware of the housing environment and its impact on design. I stayed there for a couple of years before moving to a private condominium.

I have vivid memories of the time when I was living in Toa Payoh. When we moved out from the kampong, and while waiting for the construction of our new flat, we were staying in my elder brother's 3-room flat near the construction site. During that time, my mother, brother and I would walk to the construction site to watch the progress of our flat being built, almost every morning. It was great fun. By watching the building process, we started making a connection with the flat even before we moved into it. It also happened that our neighbour in the kampong was one of the workers who were building the flat. There was a tangible human connection, and it seems that the construction was a collective effort toward building a better and new life for us all. I suspect nobody watches the construction process anymore today, which is a pity, as I think it is important to partake in this sense of co-creation of our living environment.

When we got our flat in Toa Payoh, I remember there was a big celebration. It was a major public event where everybody came to draw lots to get their flat allocation. The local officials put up tents and banners, and ceremoniously handed out the keys to the future residents. We opened the doors of the flats together with many other families in the same estate, so we got the chance to meet our future neighbours before we started living there. The whole experience also gave us a sense of anticipation, pride and ownership. For me, specifically, it was the entire experience of getting onto something new—a new home, going to a new school, buying new clothes and textbooks—a tremendously happy occasion. Sadly, today such experiences are no longer felt—everything is mechanical, being facilitated remotely by the internet—from the flat selection to the meeting with the public officials. We seem to have lost the sense of joyous celebration of moving together to a brand new neighbourhood and an altogether new living experience. Instead, we moved in with a long list of defects to be rectified and many unrealistic expectations.

These two processes of watching the construction process and celebration of a new neighbourhood are missing in today's public housing neighbourhoods—all in the name of efficiency. I believe both are important if we are thinking of building connectedness and eventually resilience. For we are not only making connections in the neighbourhood right from the start, but also helping to manage our own expectations which will reduce unhappiness with our environment later on.

3 Designing for Neighbourliness

The HDB has done much over the decades and accumulated a wealth of knowledge in planning and designing urban residential neighbourhoods. In particular, there have been significant improvements in the design of public spaces and landscaping. In the earlier years of HDB housing design, there were only green lawns, spaces for playgrounds, and open car parks. There was not much variety of communal spaces due to limited availability of financial and professional resources. In those days, we often utilised a single space for different users; for example, the open car parks were used for funerals, weddings and other events. Today, some of these uses have been accommodated either in the void decks or purpose-designed precinct pavilions. There is a variety of parks, playgrounds and other amenities. All these help to promote community use and interaction.

Over the years, the HDB has experienced different issues in the planning and design of urban neighbourhoods, as the needs and expectations of the population have evolved. In response, they have been experimenting with many solutions. Many of these strategies have been successful and could be revisited. The following examples illustrate these strategies.

3.1 Corridors

In many of the newer townships, the apartments are arranged in double-loaded corridors; we often see corridor after corridor with only solid walls and no windows at all. In the name of densification, we have created faceless corridors that would reduce the chances for neighbourly encounters.

During my stay in Toa Payoh, my family was living in an estate designed with the slab block typology—long narrow single-loaded corridors serviced the residential units. I can still remember all my neighbours who lived on the same floor, their names and their occupations. Doors and windows faced the corridor, yet, people left their front doors open, but with the grill gates shut for security. In this way, a semi-public and semi-private threshold existed between our living space and the corridor. The threshold invited neighbours to approach each other, ask for help, borrow something, share some food, or engage in other forms of interaction. It was a good local practice but fewer and fewer people are doing it today. The current design typology with double-loaded corridor configuration and planning policy play a role in discouraging this. This has been worsened by the present fire code, which disallows window openings onto the corridor. The front door must be a fire-rated door such that, technically, it must be closed at all times, which prevents us from getting to know our neighbours. So, what we have is an interrelated issue of policy, design typology, and the social consequences.

3.2 *Ground Floor Backyards*

The old HDB estates had ground floor units where people used to "colonise" their backyards, and in the process, engender a strong sense of a well-knit community and neighbourhood. I used to visit my sister who lived in a ground floor unit. The units were three to five steps elevated from the ground, giving privacy for residents who would usually leave their doors and windows open. The backyards were used for gardening, washing and drying clothes, or other social activities. My sister always cleaned and washed the backyard after doing her laundry, even though maintenance was the Town Council's responsibility. In her own way, she displayed a sense of responsibility, which comes with the default ownership.

The backyard was also a space for social encounters. Every morning, while residents were washing their clothes, the cleaners were also emptying the refuse chutes located between the backyards. This presented opportunities for the residents not only to know one another but also the cleaners with whom they formed a relationship. In the process, the cleaners did not just provide a service, but became much-welcomed members of the community. My sister always offered them a coffee and a chat. Sadly, such opportunities do not happen anymore with our new housing typologies that have no residential units on the ground floor. With the primary concern to make everything more efficient, the opportunities to have face-to-face contacts and interactions have somehow diminished. Design can inadvertently discourage connections and relationship-building.

3.3 *Distances Matter*

As residents, we usually feel safer in a neighbourhood with many people around and/or watching over things. Appropriate distances in between blocks are important—distances that are not too intrusive yet allow us to still recognise the facial features of a person who is not too far away. In Toa Payoh, I was staying in the rear bedroom looking across the carpark to the corridor of a block of rental flats. It was an interesting pastime, bordering on voyeurism, for I could almost look at what is happening on every floor; I noticed who worked late, who had just come back home, and who slept early, etc. Sometimes I could see my schoolmate walking back to his home. By observing, I had an idea what was happening in my neighbourhood; in watching the habits of my neighbours, I got to know them better. If something was amiss, I would also know, and this gave us a sense of security, and also knowledge that we could render help wherever or whenever necessary.

3.4 Elderly Residents

Planning and design of a housing neighbourhood should be flexible and accommodate development at different stages of one's life. Design should encourage people to take care of one another rather than hiding from and isolating them. I have noticed that it is a common habit of the elderly to wonder around the neighbourhood. My mother, for example, although wheelchair-bound, enjoys going to the balcony and observing the neighbours. She is still very aware of her surrounding neighbourhood, she knew people who have moved out, who have just moved into a vacant flat, etc. Someone once told me that the days of the kampong will come back when we have a lot of elderly people who are retired and hanging around in the neighbourhood like my mum. With time on hand, they would be "kaypoh" to find out who is in the neighbourhood and try to connect with them. The planning and design of a residential neighbourhood should support the concept of 'ageing in place'.

4 Planning Bidadari Estate

In our master plan and design for the Bidadari Estate, the positive experiences and the memorable moments of my younger days of living in an HDB estate have influenced my design approach. This can be seen in the following design features in Bidadari:

4.1 Ground Floor and Void Decks

In the current housing typology, the layout of ground floor is typically occupied by void decks. In Bidadari, I opted for ground floor units and integrated them with the void decks. Void decks are meaningless without people populating them. So, we planned for studio/flexi units on the ground floors opposite the void decks. With these, there is a higher chance that the families occupying the ground units would utilise the void decks more, almost as an extension of their living space. As we know from experiences, scenes of people getting together or having conversations, or of kids playing happily, all these affect the people who see it; these activities are thus positive catalysts for community building. So, planning and design must encourage these activities to happen.

4.2 Walkways

It is now quite common to include walkways in our planning, but this is often the result of rigid building arrangements. In Bidadari Housing Estate, we have planned the walkways upfront. We designed a kind of "social verandah"—a walkway that would connect major activity nodes physically and people socially. They are connected to the bus stops, the open spaces and the open greens. Almost every building block is linked across the neighbourhood.

Why the bus stop? It is a destination. People tend to linger around and chitchat while waiting at a bus stop. If there is a small vendor who sells newspapers, lighters, tissue paper, candies and snacks, people will linger a little, so we can begin to also think of the vendor as a catalyst to attract people. We have also planned for the walkways to be wide enough, not only to allowing people to pass by, but also to be able to accommodate occasional events such as a weekend bazaar, a *pasar malam*,[2] garage sale, charity event, etc. While it is up to the community to activate the open spaces by organising the right activities, we, as the architects and planners, can anticipate these activities to happen, and provide the necessary supporting design elements. Planning and design should facilitate activities and encounters to happen, and it is our role to design the infrastructure and plan the space in a strategic way in order to allow the right catalyst to come along.

4.3 Connectivity

Interconnectivity of public spaces in urban neighbourhoods is important. People have the right to access and be linked to meaningful spaces. In Bidadari, we have designed the linkages that seamlessly interconnect the housing blocks to childcare centres, senior citizens' corners, rooftop gardens and car parks. So, residents can conveniently drop their children or elderly parents, or in an event that the garden is used for a party, a service truck can readily gain access, or in a case of a funeral, a hearse can easily come into the estate. Planning must cater to such real life and day-to-day needs. A funeral, a party, and a wedding are the signifiers of important events in one's life; they bring people together, whether it is in moments of happiness or sadness. So, as architects and planners, we have to design and plan urban spaces in order to allow for these events to happen easily and flexibly (see Fig. 1).

5 Planning Future Resilient Communities

The fundamental principles of planning for a resilient community are as follows: first, planning and design must bring people together; and second, it must encourage people to know one another; from thence, everything else will follow. It is important to know that the design features could only suggest potential usage of the spaces,

---

[2]An Indonesian and Malay word that literally means "night market". It is a street market found in Singapore and other parts of South-east Asia that operates only in the evening, usually in residential neighbourhoods.

Fig. 1 The Social Verandah that links different neighbourhood. *Source* MKPL Architects, Singapore

but we need "champions" to demonstrate, for example, what living and taking care a neighbour means and how ageing in place with a 4-Room flat is possible.

In master planning, it is difficult to guess the demographic profile and behaviour of the people who would eventually live in the estate. However, there are certain criteria in the design brief that can give us clues. For example, when the brief mentioned about designing studio units, there is already a defined set of criteria for people who qualify to stay in those units, and we would then know the likely profile of the future residents. Then, there are established space standards that would suggest how people would live and what sort of communal spaces are needed to help them not to feel isolated. There are also existing studies or research that would suggest certain population numbers that make for a comfortable community. With this information, we can then begin to create and plan the right cluster sizes so as to encourage neighbourliness and community building.

5.1 Flexibility

There could be many ways of anticipating the type of people who would eventually stay in an estate that is being planned, what their needs might be, and how they are likely to behave especially with regard to the various land uses and urban spaces. However, questions for example, of whether there would be more working population or more residents staying at home, do not really matter. The important consideration is that the design must be flexible enough to be able to cater to people who might live there from a young age until they grow old. The planned and designed urban spaces should allow for various activities and events to happen. In Bidadari, for example, due to the undulating terrain, we have designed gentle ramps for barrier-free access throughout the estate. It is inclusive and can accommodate elderly wheelchair users, kids with tricycles, parents pushing prams, and many others. The planning and design of the neighbourhood should be future-oriented to ensure that there is sufficient flexibility that allows for the community to grow old within the neighbourhood. The flexibility might even be embedded into the internal layout of the units. For example, our design allows two studio units to have a connecting door for sharing with friends or with the caregiver while maintaining privacy for both units (see Fig. 2).

5.2 Ageing-in-Place

In an NUS architecture studio on design for ageing, one tutor commented '*why can't we imagine that life will only get better when we turn seventy, and not worsen*'. It made me ponder.

On reflection, perhaps he was right! Typically, we work hard when we are young, and when we grow old, we often find that we cannot enjoy life as much as we would like to. And, the living options for ageing are presently quite limited, for example, nursing homes. If we view this from the family life cycle perspective - as parents always work hard for their children, so, when they grow old, the children who would have had a better life can, in return, make sure their parents can retire gracefully. If we apply this cycle to society as a whole, we will have a positive virtuous cycle, and society would be more stable and resilient. Ageing becomes a natural order of things. So, in an economic system, we should spread our resources to cater for the various stages of family life.

With specific regard to planning, in the early days of public housing design in Singapore, the government distributed the financial resources carefully to conserve them. The HDB, for example, placed the lifts to stop at the staircase landing in order to service two levels with one landing. However, this approach is no longer tenable for an ageing society. The lifts must now be upgraded to address the needs of different age groups.

On a macro level, when the country is young, the government may allocate financial resources to other priorities, for example, education, infrastructure, or others. But, as the population ages, we need to recalibrate and reallocate the resources to give proper attention to issues that come with an ageing society. The new questions that arise include the following: *what are the living options for the ageing population? How do we change people's perception about providing facilities for the ageing population? Why must a nursing home be a standalone entity and not an amenity naturally embedded into an urban neighbourhood? If childcare centres can be treated as part of neighbourhood amenities, why cannot we think of nursing home facilities in the same way?* These are relevant questions, given that in Singapore, family ties are still strongly valued, and the notion of planned nursing home facilities

Fig. 2 Flexibility in 4-Room dwelling unit layout. *Source* MKPL Architects, Singapore

for the aged away from residential neighbourhoods, as commonly found in affluent western societies, is something that needs careful evaluation. The design solution is to address these issues upfront during the planning stage, rather than relocating the elderly population out the community and isolating them in standalone facilities when the time comes.

6 Planning More Integrated Development

The present and future challenge in housing development is to change the mindset of government agencies, planners and architects, in terms of not to think of housing, healthcare, and eldercare as separate entities, but plan them as one ecosystem and maintaining the sense of an integrated neighbourhood.

At present, the government is moving towards integrated development gradually. This concept of "active ageing" is in response to the expected challenges in future where the characteristics of the elderly population would be quite different from the present. The future elderly population would probably not be happy with just sitting around a chess board and drinking, but would be healthier, physically more mobile than today's 70- or 80-year olds. They are likely to be more entrepreneurial. We have to accommodate the diverse needs of this population segment with careful planning and design.

In neighbourhood planning and design, architects and planners are typically given a brief from the HDB to consider various facilities that will eventually be operated by several government agencies. These agencies are usually allocated spaces that are designated to fit in the various facilities. However, the actual engagement with the respective agencies could only happen during in the construction stage, which may be an opportunity lost or one that comes too late. For effective integration of the various facilities in the plan and design of urban neighbourhoods, collaboration with such agencies and stakeholders should start right from the planning stage.

There are many ways of achieving better integration, as follows.

(a) ***Connectors***: In the nursing home project—the Bukit Batok Nursing Home—which our firm undertook, our task was to design the nursing home to be embedded within a public housing estate. The site is surrounded by public housing blocks on three sides. Between the housing blocks and the nursing home is a neighbourhood park, and on the opposite side across the street is a primary school (see Fig. 3).
Initially, we had many design ideas to integrate the nursing home with the adjacent neighbourhood; for example, the nursing home can share car parks with the public housing, or the rooftop garden of multi-storey car park can be connected to serve the nursing home. Unfortunately, the timing did not go well with the planning, as HDB had already planned and designed the neighbourhood and proceeded with its construction. Only the plot for the future nursing home has been left empty. Hence, the integration could not happen seamlessly.

Fig. 3 Location plan of the Bukit Batok Nursing Home. *Source* MKPL Architects, Singapore

Nonetheless, we sought to find other opportunities for better integration. We designed the building to be a connector within the neighbourhood. Hence, when residents are on their way to the neighbourhood school, after passing the park and crossing the road, they will naturally pass by the nursing home. Thus, by integrating the nursing home into the residents' daily routine, we hope that they would begin to wonder about the facility and get interested in its activities. This way, the nursing home would not be alienated and isolated from its neighbourhood and the local community. In similar vein, we designed the ground floor to be porous and even suggested bringing in a catalyst—the ubiquitous *ice-cream vendor*, a familiar figure from bygone days.

So, as architects and planners, we can influence people's perception through better neighbourhood design. We can situate a building in a strategic location, imagine the type of activities that may happen, design the spaces to allow them to happen, and leave the rest to the facility operator and the community.

(b) **"Frontage" and co-location**: In the Bedok Integrated Public Housing Project, our firm was tasked to design an 'active ageing hub' within a residential neighbourhood. In our plan, instead of hiding facilities for the elderly residents, we created them as a "frontage"—we celebrated its architecture in the most prominent location we can think of. We integrated the amenity by situating its facilities along the corridors leading to the housing blocks. This way, people do not have to make a special effort to look for these facilities; instead, they become very much part of the everyday life of local residents (see Fig. 4).

Co-location helps in creating opportunities for interaction. For example, we inserted a communal kitchen triangulated with the active ageing hub and a void deck. So, on festive or special occasions, people can cook at the kitchen when using the void deck as a gathering space. By integrating the facilities, we facilitated encounters so that people can build relationships. The community can also relate to the active ageing hub in a more personal and proactive manner. Overall, people would begin to feel less intimidated by the facility for the elderly residents. In this way, we contributed, as architects and planners, by using good planning and design to break down barriers, mitigate social stigmatisation and head-on tackle the *Not In My Back Yard (NIMBY)* syndrome.

(c) **Shared spaces**: In another example, we co-located a childcare centre and senior care centre by designing a shared common veranda outside. We observed that parents are usually hanging around the common area outside a childcare centre to watch their kids, especially during the first few weeks of school. Thus, the co-location not only accommodates families who can send both their children and elderly parents together to the co-located facility but also allow opportunities for the elderly to look over the children or interact in a meaningful way.

7 Multi-stakeholder Collaboration

Integrated development requires effective multi-stakeholder collaboration, particularly at the working level. Currently, there are still roadblocks to such collaboration; it is still not an entirely seamless process. The main challenge is the existing mind-set of many of the agencies, with each one having its individual agenda. For seamless collaboration, we need a common vision and outlook amongst the various agencies that have a stake in the project. For example, in the nursing home project in Bukit Batok, the challenge in integration was the strict roles and boundaries adopted by the participating agencies. Often, the ground staff tend to adhere strictly to the boundaries laid down in maintenance contracts, so as to avoid problems such as which agency should pay and which should maintain particular segments of a facility.

We need an agency that keeps its focus on the common agenda. This agency could facilitate seamless connectivity, better use of amenities, landscaping and land use, and ensure the creation of soft boundaries between adjacent sites. The ultimate goal is for the welfare of residents, with each participating agency contributing to meeting the common agenda rather than individual organisational objectives. For now, there is still work that needs to be done, and it will take time and effort, especially when it comes to implementation on the ground.

The barriers and solutions can be clearly illustrated in the planning of Bidadari Housing Estate, where at the master plan level, there were various decisions that had to be made collaboratively. Two examples are discussed here for illustration.

Fig. 4 Co-location and active frontage of Bedok Integrated Public Housing Project. *Source* MKPL Architects, Singapore

First is the location of the current Upper Aljunied Road. In the original planning, HDB and the Land Transport Authority (LTA) were keen to maintain the current location of the road; there were two rows of large rain trees along the road that NParks wanted to retain as well. However, it was later realised that the road had to be widened to serve one of the new neighbourhoods. *But, how can we widen the road without cutting the trees?* Moreover, keeping the current location of the road means separating the proposed lake from the future Bidadari Park. But, as we began to understand the situation better, we were able to propose shifting part of the Upper Aljunied Road down and connect it back with the tail end of the existing road. By doing this, we were able save some valuable trees, maintain the connection between the existing park and the planned lake, and convert the remaining portion of the road to a pedestrianised heritage walk (see Fig. 5).

Second is the landscaping of the housing neighbourhood in this estate. The landscaped areas within residential neighbourhoods are mostly owned and managed by the Town Council, whereas roadside greenery is owned and managed by NParks. Every agency in Singapore is aware of the fact that if a facility or area is beyond its boundaries, the agency is not responsible for maintaining it. This segregation creates a situation where there is inconsistency in the landscape character throughout a neighbourhood, largely due to the unequal technical knowledge and expertise of different agencies. But, we think and ask: *Why should the landscapes be different?* In Bidadari, we advocated for consistent landscaping throughout to create a distinct characteristic for the township, and for the residents to feel that they are indeed living in a park. As an architect, I would advocate for NParks to be a national agency for all open spaces and landscaping.

Fig. 5 Planning for Parks in Bidadari Housing Project. *Source* MKPL Architects, Singapore

In pursuing these various projects, I realised that an architect alone has very limited powers to make significant changes. Obviously, the best and most effective approach is to convince and change the mind-set of the various agencies, for they are the agents of change. But, my modest role and contribution could only be to engage and convince the people in the agencies whom I worked with so that they see the goodness and benefits of a design. When you influence more and more people, you will have a critical mass of people who become willing to change things. Sometimes when a project is completed, I would invite the people concerned to the site to revisit how an idea has been implemented in reality, observe how people use the space, and validate the initial design vision. I always feel that every project that pushes for new ideas will help in the next project. It emboldens people's mind and enables them to move forward.

8 Final Reflections

In the preceding paragraphs, I have pondered, as a practising architect and one who has grown up in a HDB neighbourhood, on how we can design and build a more resilient community? The strength of social bonds in an urban neighbourhood can only be formed when the neighbourhood is planned for people to meet and know one another. 'Knowing' one another is the most basic and fundamental building block of a neighbourhood. Neighbours can help watch over our house, keep our newspapers from piling up, or water our plants when we are away, for instance. These simple acts are but signs that neighbours are helping each other. By knowing and understanding one another, many things can happen, and over time, we can build community self-reliance and resilience. Thus, architects and planners have to consciously design spaces to allow opportunities for people to meet and know each other. In the process, we can facilitate and help build community self-reliance and resilience. When adversities happened, the community would come together and overcome them collectively.

Finally, empathy in design. Since I graduated from architecture school thirty years ago, I have learned that an architect needs to embrace and empathise with the diverse points of views from other disciplines. The architect has a unique role, which is to synthesise different issues and seek opportunities from various disciplines to physically envision the possible solutions.

An architect needs to have huge empathy for the end users as well. Empathy will help us understand how people could utilise the space and make provisions for it to happen. For example, in designing for the ageing population, we need to have plenty of empathy for both residents and caregivers. In our nursing home project, I have insisted that we have to design a nice environment for the caregivers as well. It is not always easy, as architects need to decide amongst different solutions and prioritise them into workable outcomes. But, at the end of the day, to me, the resolution must come with a design that uplifts user experience. It should be as natural as possible, and should appear to be effortlessly and seen as part of the daily life of the people who use it.

Printed by Printforce, the Netherlands